ENERGY USE
The Human Dimension

ENERGY USE
The Human Dimension

Paul C. Stern and Elliot Aronson, Editors

Committee on Behavioral and Social Aspects
of Energy Consumption and Production

Commission on Behavioral and Social Sciences and Education

National Research Council

W. H. Freeman and Company
New York

Notice: The project that is the subject of this report was approved by the Governing Board of the National Research Council, whose members are drawn from the councils of the National Academy of Sciences, the National Academy of Engineering, and the Institute of Medicine. The members of the committee responsible for the report were chosen for their special competences and with regard for appropriate balance.

This report has been reviewed by a group other than the authors according to procedures approved by a Report Review Committee consisting of members of the National Academy of Sciences, the National Academy of Engineering, and the Institute of Medicine.

The National Research Council was established by the National Academy of Sciences in 1916 to associate the broad community of science and technology with the Academy's purposes of furthering knowledge and of advising the federal government. The Council operates in accordance with general policies determined by the Academy under the authority of its congressional charter of 1863, which establishes the Academy as a private, nonprofit, self-governing membership corporation. The Council has become the principal operating agency of both the National Academy of Sciences and the National Academy of Engineering in the conduct of their services to the government, the public, and the scientific and engineering communities. It is administered jointly by both Academies and the Institute of Medicine. The National Academy of Engineering and the Institute of Medicine were established in 1964 and 1970, respectively, under the charter of the National Academy of Sciences.

Library of Congress Cataloging in Publication Data
Main entry under title:

Energy use.

Bibliography: p.
Includes index.
1. Energy consumption—United States—Psychological aspects. 2. Energy industries—Social aspects—United States. I. Stern, Paul C., 1944-
II. Aronson, Elliot. III. National Research Council (U.S.). Commitee on Behavioral and Social Aspects of Energy Consumption and Production.

HD9502.U52E565   1984   306'.03   84-1633
ISBN 0-7167-1620-8
ISBN 0-7167-1621-6 (pbk.)

Committee on Behavioral and Social Aspects of Energy Consumption and Production

Elliot Aronson, *Stevenson College, University of California, Santa Cruz* (Chair)

Robert Axelrod, *Institute of Public Policy Studies, University of Michigan*

John M. Darley, *Department of Psychology, Princeton University*

Sara B. Kiesler, *Department of Social Science, Carnegie-Mellon University*

Dorothy Leonard-Barton, *Sloan School of Management, Massachusetts Institute of Technology*

James G. March, *Graduate School of Business, Stanford University*

James N. Morgan, *Survey Research Center, University of Michigan*

Peter A. Morrison, *The Rand Corporation, Santa Monica, California*

Lincoln Moses, *Department of Statistics, Stanford University*

Laura Nader, *Department of Anthropology, University of California, Berkeley*

Steven E. Permut, *School of Organization and Management, Yale University*

Allan Schnaiberg, *Department of Sociology, Northwestern University*

Robert H. Socolow, *Center for Energy and Environmental Studies, Princeton University*

Thomas J. Wilbanks, *Energy Division, Oak Ridge National Laboratory, Oak Ridge, Tennessee*

Sidney Winter, *School of Organization and Management, Yale University*

Paul C. Stern, Study Director
Richard Hofrichter, Research Associate
Ellis Cose, National Research Council Fellow

Preface

In 1980, at the request of the U. S. Department of Energy, the National Research Council established the Committee on the Behavioral and Social Aspects of Energy Consumption and Production. Our charge was to undertake a broad review of literature in the behavioral and social sciences with potential relevance to an understanding of energy consumption and production in the United States. The need for such a review was pressing. To take one rather dramatic example, violence broke out among motorists at gasoline stations during the 1979 petroleum shortfall. That behavior was not anticipated by policy makers, although the basis for anticipating it was available in the data and ideas developed over the years in social psychological research on the antecedents of aggression, as well as on ways of curtailing its harsher consequences. The effects of a far greater shortfall could make the violence of 1979 look pallid by comparison unless U. S. policy makers have at their disposal a sophisticated understanding of such phenomena.

By establishing this committee, the Department of Energy has explicitly recognized the potential of the noneconomic social sciences for informing energy policy. This was clear from the initial description of the committee's purpose and tasks:

> Economic paradigms, together with assessments of the potential contributions of new and existing technologies, will continue to provide the basis for the analysis of alternative public policies relating both to energy production and consumption. At the same time, there is considerable evidence to suggest that the noneconomic behavioral and social sciences can contribute significantly to such analyses in at least four major areas.

"1) *Increasing the accuracy of predictions of behavioral responses to economic incentives.*" For example, analysis of social, psychological, and organizational factors mediating energy use can help explain why economic incentives have not always had the expected effects on the behavior of individuals and firms. Research in the noneconomic social sciences can also illuminate the processes of formation and change of consumer preferences, which are usually treated as exogenous in economic models.

"2) *Increasing our knowledge of noneconomic approaches to behavior change.*" It has long been observed that under some conditions, changes in social norms, such as those encoded in civil rights laws, produce corresponding changes in behavior. But it is also true that under other conditions behavioral changes occur in the absence of legal sanctions, economic incentives, threats, penalties, and the like. In recent years, research in social psychology has made great progress in differentiating among those sets of conditions. For example, much research has shown that when compliance is induced with minimum pressure, the resulting changes in behavior can be far stronger than those induced by strong sanctions. Such processes of social influence are almost certainly applicable to energy consumption.

"3) *Increasing the capacity of the public to make choices about new and existing energy technologies.*" Knowledge developed by psychologists, sociologists, and other researchers concerned with organizational behavior can be useful for improving communication systems, understanding the basis of public perceptions of risk from energy technologies, and understanding the responses of producers to changing consumer preferences.

"4) *Anticipating the consequences of alternative energy policies.*" Energy policies may, for example, influence interregional migration, create needs for new types of skills and training within the work force, and alter housing and transportation patterns. Such changes in major social systems and processes may potentiate or undermine the intended effects of policies, or they may produce important secondary, unintended effects.

Thus, significant potential contributions from the noneconomic behavioral and social sciences are evident. But these fields have only recently begun to address energy issues. As a result their contributions can at present only occasionally take the form of propositions derived specifically from the empirical study of energy issues. More often, basic knowledge developed through the study of behavioral and social processes can call attention to opportunities or problems that might otherwise be overlooked by energy policy makers. Although it cannot be used to prescribe the details of implementation, this knowledge can suggest strategies for implementing policies and programs; it may point toward new policy initiatives; it may also suggest ways to acquire knowledge that could reduce the likelihood of major policy mistakes.

This book draws together these contributions as they relate to a selection of energy policy issues. Because of the wide range within which potential contributions exist, we have chosen to explore only part of the territory. We treat selected contributions in depth rather than give a superficial account of many, so that readers—especially those unfamiliar with work in the noneconomic behavioral and social sciences—may gain a more substantial understanding of how the knowledge and concepts developed in these disciplines can be useful in energy policy debates.

Our goal, in short, is to offer relevant ideas, theories, and research results from several social and behavioral sciences in the hope of increasing the understanding of policy makers and citizens about vital factors that affect some of the ways people relate to the energy system. It is our belief that without such understanding, energy policy will be at best incomplete, and, occasionally, misdirected.

Finally, I should say something about how our work was actually done. We met periodically as a full committee discussing ideas, reviewing literature, arguing, criticizing, honing, and sharpening written drafts. The committee meetings themselves were almost always stimulating; sometimes they were even enjoyable! In a real sense, it can be said that every member of this committee is an author of the entire manuscript because each of us contributed ideas, insights, criticisms, or research to every chapter.

In addition, we divided into subgroups for the actual drafting of chapters, so that some of us are more responsible for certain chapters than for others. It goes almost without saying, however, that no book or chapter can be effectively pulled together and synthesized by a committee—or even by a subcommittee—unless someone is willing to take on an unusual degree of responsibility for major segments of the project. We were fortunate to have more than a few committee members who performed this function—each for his or her major segment. Paramount among these is our study director, Paul C. Stern, who did the lion's share of the cajoling, synthesizing, and editing. I am especially grateful to him for his wisdom and effort.

In addition, it is a pleasure to express my personal appreciation to several people outside the committee who contributed a great deal to the project. These include William Lewis, Lester Silverman, Nicolai Timenes, Peggy Davis, Diane Pirkey, and Barry McNutt of the Department of Energy; David Goslin and Heidi Hartmann, executive director and associate executive director of the Commission on Behavioral and Social Sciences and Education; Richard Hofrichter, our research associate, and Ellis Cose, our NRC fellow.

We would also like to acknowledge editorial help provided by Eugenia Grohman, the Commission's associate director for reports, and Ellis Cose; and secretarial help provided by Wendy Siniard, Andrea Gershenow, Grace Stewart, and Donna Reifsnider. Various chapters were reviewed by a great many individuals, alas too numerous to name.

Elliot Aronson, *Chair*
Committee on Behavioral and Social Aspects
of Energy Consumption and Production

Contents

ENERGY USE
The Human Dimension

1

The Human Dimension

In March 1982, when we started writing this book, a headline in the *New York Times* announced, "Energy Shortage Eases Materially." The article went on to quote the then U.S. Secretary of Energy (in Martin, 1982): "I'm not going to say the energy crisis is over, but we are certainly heading in the right direction." More recent news stories continue to echo this message. In a more recent public opinion poll, only 7 percent of the respondents considered energy one of the nation's two most pressing problems (Hershey, 1982). Had the energy crisis really ended?

It would be comforting to believe that the energy problems of the United States are well on their way to solution and that a new analysis of the national energy situation is unnecessary. Unfortunately, the data do not support such an outlook. Recent changes in world oil markets do not herald the end of the nation's energy problems. In spite of temporary surpluses of oil, the United States still imports about 20 percent of the oil it uses from the Middle East—an area that has certainly not grown more stable in recent years. Furthermore, the United States has agreed to share oil with its allies in case of a shortfall, and some of these allies are critically dependent on Middle Eastern oil. Thus, a serious disruption in the supply of oil from the Middle East, which is always possible, would have grave consequences for the United States. As James Schlesinger has said (quoted in Martin, 1982), "The energy crisis is over until we have our next energy crisis."

Granted that a combination of increased energy production and conservation to meet the nation's needs is still an urgent national priority, why another study? The answer is that most previous analyses of the conservation and production of energy have all but ignored an important aspect of the situation—what we call, for lack of a better term, the "human dimension." The human dimension refers to the rich mixture of cultural practices, social interactions, and human feelings that influence the behavior of individuals, social groups, and public and private institutions. To consider the human dimension is to recognize that the behavior of individuals and institutions is multiply determined.[1] This may seem an obvious point, but it has important ramifications.

Most analyses proceed from the simplifying assumption that energy producers and consumers are rational economic actors: that is, that they are motivated to maximize the value of some objective function, such as income, profit, or organizational size. Individuals and organizations are assumed to behave as if they had carefully calculated their self-interest and acted accordingly. This assumption is a useful simplification. It accurately predicts, for example, that when oil prices rise relative to the prices of other fuels, some energy producers will invest in oil exploration, and some energy users will switch to other fuels or purchase more energy-efficient equipment. But such aggregate truths conceal great variation among energy producers and users, and some of that variation can be understood in terms of other concepts and analyses. People have values, dreams, and social needs, and they sometimes act on them. They often act out of habit, laziness, duty, trust, or a desire to please others, and they act differently than they would if they were to carefully calculate their self-interest. People also form organizations, families, political parties, and social movements, and these social groups, like individuals, are more than rational economic actors. Groups, organizations, and governments often follow routine, precedent, ideology, or the example of a leader rather than act on careful calculations. Their choices among alternatives certainly are influenced by analyses of expected costs and benefits, but they are also influenced by other factors: the outcome of internal political struggles, the recent choices of similar groups, the desire to promote socially shared values, and the personal preferences of individuals in powerful positions. Social groups maintain coalitions out of tradition, build monuments for prestige, and fight battles for honor.

After the fact, many of the actions of individuals and groups can be interpreted as if they were the result of rational calculations of self-interest. But for policy purposes, it is crucial to be able to predict, as well as to interpret, the behavior of individuals, groups, and institutions. It is also essential to have a wide range of policy alternatives available for debate and adoption as energy conditions change.

SURPRISES IN THE ENERGY SYSTEM

Several recent events demonstrate that there is a pressing need for improved prediction as well as for new policy options for energy. In the past, analysts have frequently been surprised when well-engineered energy technologies fail to work as expected, or when carefully planned policies or programs are greeted with public apathy or opposition, or when energy users behave very differently from what was predicted or expected. Often the surprise is traceable to the fact that the analysts had not paid enough attention to crucial processes in individuals, organizations, or social institutions. There is a body of empirical knowledge about individual and social behavior that can help avoid or cope with such surprises, but this knowledge has been largely ignored by policy makers.[2] Only a few examples are needed to illustrate how, by taking the human dimension more fully into account in energy policy, the nation could render the prospect of energy crisis less threatening, make important energy production technologies more reliable, and enable the public to make better informed choices about energy use.

Responses to Oil Shortages

In 1979, a minor shortfall in oil supplies led to widespread hoarding of oil products, long lines at gas pumps, the installation of dangerous extra fuel containers in private cars, and even occasional violence in the form of fistfights and shootings. The energy policy community responded with proposals designed to keep gas lines from forming as a way of preventing such unexpected and antisocial behavior in the next period of shortage. But these proposed alternatives were based more on reactions to a politically unacceptable situation than on a careful analysis of behavior in crisis situations. Such reactive policy making runs the risk of substituting a new difficulty for the old one. For example, one way to make shortages disappear is to allow prices to rise. This could eliminate gas lines by making it prohibitively expensive for some people to buy gasoline. Under the conditions of 1979, such a policy may have been better than what actually occurred. But if this policy were practiced in a serious shortfall, the frustration and deprivation that would follow uncontrolled price increases might greatly increase other undesired effects, such as the siphoning of gas from gasoline tanks, vandalism or robbery of gasoline stations, or random acts of aggression. A policy of offsetting price increases by taxing the increase and recycling the tax revenue would prevent deprivation only if the recycled revenues were to reach the neediest segments of the population very quickly and effectively. So far, these issues seem not to have been carefully considered.

Reactive policies fall short in a more fundamental way as well. They are based on implicit acceptance of an energy system that is organized to make planning for gasoline emergencies a necessity, rather than on a view that the energy system might be restructured, for example, to decrease the dependence on automobiles, the prime users of gasoline.

The human dimension of energy is important both in understanding responses to crises and in organizing the energy system to be less crisis-prone. For a number of years, research in the social and behavioral sciences has been analyzing human behavior in crisis situations. The social processes involved—panic behavior, emotional reactions to threatened loss of freedom, and similar types of responses—are not easily understandable in terms of rational choice, but they can be understood in terms of concepts developed by psychologists and sociologists (e.g., Schultz, 1965; Brehm and Brehm, 1981). Other research findings and insights can be used to inform political debate on the issue of preventing energy crises. These include research on the ways settlement patterns, land-use policies, and other social forces affect energy needs. In addition, the vision that has limited policy makers to a reactive approach is in itself the product of a fundamentally important process—one that shapes the way people think about energy. We discuss this process in Chapter 2.

Three Mile Island

The accident at the Three Mile Island nuclear power plant focused attention on the human dimension, but this time in an area of energy production. The President's Commission on the Accident at Three Mile Island concluded that ". . . the fundamental problems are people-related problems and not equipment problems." A major problem was that the plant's control panels had been designed in a way that confused operators when the equipment malfunctioned. This confusion could have been predicted, given knowledge of the plant design and of processes of human perception and information processing. Both perception and information processing have been studied extensively (e.g., Haber, 1969) and have been the subject of considerable applied research with airplane control panels (Roscoe, 1980) and other technologies (e.g., Sheridan and Johannsen, 1976; Van Cott and Kinkade, 1972). Such behavioral research has been used in both the public and private sectors for many years—since long before the development of a nuclear power industry—and has led to improved training of equipment operators and more human-centered design of machines. It seems clear, however, that this expertise had not been used in the nuclear energy industry. Since the disaster, considerable new investment has been made by the Nuclear Regulatory Commission and the industry in understanding how human beings interact with the equipment and in developing more understandable equipment design (Mynatt, 1982).

Another major problem at Three Mile Island was a lack of coordination among relevant personnel. For example, the "incident" was not reported by operators to their supervisors, or by utility officials to federal and state authorities, for some time. This delay occurred in spite of company policies and federal regulations requiring immediate reporting. However, there are reasons that employees do not always follow company policies and that organizations do not always follow federal regulations. Reporting a power plant accident may invite expensive consequences for a utility company even if the incident is insignificant, so both operators and utility officials hesitate to report as significant problems what may turn out to be unimportant events. Because minor events are much more common, it is not surprising that operators and officials may wait in the hope that a situation will resolve itself.

At Three Mile Island, in fact, the failure of responsible personnel to manage events effectively and to report the accident quickly seems to have been partly due to the reward structure of the workplace (Egan, 1982). There is a considerable body of research on organizations as social institutions that sheds light on the processes that can operate in organizations to undermine the intent of stated policies (e.g., Cyert and March, 1963; Kaufman, 1971, 1977; March and Olsen, 1982; Sproull et al., 1978). An informed expert in organizational behavior would probably have been able to see that the existing regulations were inadequate to ensure reporting of the events at Three Mile Island or similar reactor accidents. With available knowledge, the society could probably develop a more effective oversight procedure or change the balance of incentives to favor more rapid reporting of potential disasters.

Nonresponse to Energy Audits

Free or low-cost home energy audits are being offered by many utility companies around the country. The typical level of response is less than 5 percent over the duration of a program (Hirst et al., 1981; Rosenberg, 1980). In California, for example, electric and gas utilities publicize their free energy audits both in the media and in bill inserts, and the publicity assures the consumer that an audit would be beneficial in terms of both savings and comfort. By 1982, however, only about 2 percent of the eligible California customers had taken advantage of the offer.[3]

Why do people pass up information that is both useful and free? There are several possible reasons. They may not trust the utility as a source of information; they may not believe the offer of something for nothing; they may be unable to arrange a convenient time for the energy audit to take place; they may not even see the announcements of the program. Or they may be unwilling or unable to act on information about energy-saving investments for their homes because they cannot do the necessary work themselves and believe that they cannot find a reliable contractor, and so may see the information as irrelevant.

If energy audit programs are to be of any substantial practical value to residential energy users, they must be based on improved understanding of consumers' beliefs and motives relating to home energy conservation. Understanding the communication processes operating in this situation would also be useful. However, energy audits are usually seen by policy analysts as an almost irresistable offer of something for nothing rather than as a process involving communication, motivation, and belief. Much relevant research has been done on communication processes, including some studies of energy information as communication (e.g., Ester and Winett, 1982; Stern et al., 1981), but the findings have rarely been integrated into the design of energy audit programs.

These examples illustrate areas in which knowledge of the human dimension can be used to improve prediction and make energy policies and programs more effective. But this knowledge has not yet been systematically developed, largely because the social and behavioral sciences that offer insight into the human dimension have not been called upon to address energy policy issues very often or very intensively.[4] It is therefore necessary for this book to be more of an exploration than a prescription. By reviewing the human side of a sampling of energy issues that relate primarily to energy use, we hope to add a needed dimension to thinking about energy policy and to offer new ideas for tackling a few difficult energy problems.[5]

About This Book

For our study, we selected for analysis three energy issues we consider to be among the most important facing the society:
1) the nature and determinants of energy consumption;
2) the problem of preventing, preparing for, and responding to energy emergencies; and
3) the attempt to meet energy needs through action at the local level.

This book considers how policy in each of these areas could be improved by using knowledge of the entire range of factors that influence individual and social behavior. The central purpose is to improve energy policy by broadening analysis to incorporate the human dimension. The next chapter provides an essential context for this endeavor. It offers an important example of how patterns of approaching energy issues have restricted energy policy options and how a broader view can open new possibilities for debate. In a modern society, energy has at least four distinct aspects: it is simultaneously a commodity, an ecological resource, a social necessity, and a collection of strategic materials. But in thinking about energy, people tend to focus on only one of these aspects at any one time. This narrow focus is one reason that general and expert opinion about energy has been so volatile and that policies often seem inappropriate by the time they are implemented. The dominance in energy policy analysis of one view of energy—that it is essentially a commodity—helps explain the typical conflict over energy issues and the failure of the political system to even debate many plausible energy alternatives. Chapter 2 thus sets the stage for a broader analysis of some specific energy policy issues.

Chapters 3, 4, and 5 analyze present energy consumption and the potential for improved efficiency in energy use. Energy users act in an environment that has some characteristics of a market (for example, fuels are available for purchase at a price), but in which other powerful forces also operate—usually to restrict consumer choices. Thus, the market metaphor is incomplete and even misleading at times. For example, the actions of intermediaries, such as manufacturers, designers, and builders, limit the choices available to the ultimate energy user. Within these limits, energy users are presented with conflicting, confusing, and often untrustworthy information about energy alternatives and are faced with an uncertain future. Furthermore, they cannot conveniently monitor the amount of energy they use nor easily evaluate the success of their efforts to modify consumption.

Most energy users are in a situation like that of customers in a supermarket where nothing is marked with a price except the bottom line of the cash register receipt. In such an environment—where it is very difficult or impossible to know the costs of individual purchases—the requisites for rational choice are not present; other processes have major effects on behavior. People will respond by following routine, borrowing ideas from neighbors, maintaining habits, and avoiding what are seen as untrustworthy sources of information. Social and behavioral scientists, particularly psychologists and sociologists, have investigated such processes, both in general and in relation to energy consumption, and their findings are examined in the context of energy. Chapter 3 focuses on the environment of energy consumption; Chapter 4 considers how individual and household energy users behave in this environment; and Chapter 5 examines organizational factors as they affect energy use by organizations and as they influence the actions of energy intermediaries. The implications for policy and programs affecting energy consumption are also discussed.

Chapter 6 examines the problems of preventing, preparing for, and responding to energy emergencies such as might be caused by a major disruption of oil imports, a regulatory decision to shut down nuclear power generation, or other disruptive events. A broad view of these problems highlights the ways emergency preparedness and prevention are interrelated, the linkages between emergencies, and the likelihood of social conflict in emergencies. A limited but useful empirical basis for understanding social preparedness and response in energy emergencies exists in the findings of social science research on responses to natural disasters, individual risk-taking, stress responses, and other related subjects. Research suggests, for example, that beliefs about who is responsible for an emergency and expectations of government action may have major effects on social response. Research also shows that carefully constructed contingency plans often remain unused in a crisis. However, such research also suggests that changes in the planning process could make responsible officials better prepared for a crisis and could increase the public's willingness to cooperate with official contingency plans.

Chapter 7 discusses energy activities at the local level. Over the past decade, thousands of local governments and community groups have taken action to provide for the energy needs of the people and institutions they serve. Although this phenomenon has not been seriously studied, it is significant, because local action holds the promise of improved energy management both under ordinary conditions and in emergencies. Local action is also attractive because it may give communities an increased measure of control over their destinies and because it may make the national energy system more flexible and resilient. Some analysts believe this promise can be realized with appropriate policies. Others believe that the political and other difficulties that have beset national energy policy are equally serious at the local level. They believe that a focus on local action will be a wasted effort. Not enough is now known to justify a choice between these views, but the capability to gain valuable knowledge from recent experiences exists. Chapter 7 identifies a number of social and political processes that should be carefully studied because they may be critical for understanding the outcome of local energy efforts and the potential of localism as a national energy strategy.

Chapter 8 presents the committee's conclusions and a summary of recommendations. It shows how new ways of thinking about energy policy can grow from clearer recognition of major features we identify in the energy system: the prevalence of uncertainty and mistrust; the perennial issue of control; the great diversity of needs, conditions, and sources of influence; and the various predictable but "nonrational" elements of individual and organizational behavior. Such recognition leads to recognizing energy problems and solutions as located within complex social systems, to considering an adaptable energy system as an alternative to one based on better planning for future events, to emphasizing the processes as well as the outcomes of policy development. Such shifts of outlook can bring promising new policy options into focus and broaden the range of alternatives for public debate.

This report is not a comprehensive study of the human dimension of energy; our intention is rather to underscore the importance of the human dimension by analyzing selected topics. For some of the topics we do not examine, overlooked social institutions and social or psychological processes play a significant role. Some topics have been the subject of extensive empirical research by social and behavioral scientists; while for other topics such research is needed. The following topics, while not covered in this report, are also part of the human dimension of energy:

Energy Production Processes.   Energy production has social and psychological aspects that are often overlooked—as the Three Mile Island accident amply demonstrated. Organizational, political, and decisional processes affect energy production in other ways as well: examples include political choices to support particular energy production technologies and corporate choices about high-risk energy investments.

Social Implications of Major Energy Technologies.   There is an extensive literature on "social impact assessment" (e.g., Finsterbusch and Wolf, 1981), as well as available research on particular related topics, such as the adaptation of individuals and social institutions in energy boomtowns (e.g., Freudenberg, 1982) and the social effects of power-plant-siting decisions. The subject is also alive in the policy world: a 1983 decision of the U.S. Supreme Court, which overturned a 1982 appellate court decision, gave the Nuclear Regulatory Commission permission to restart the undamaged nuclear plant at Three Mile Island without first evaluating possible effects on the psychological well-being of people in the surrounding communities, who live in fear of a recurring catastrophe.

Relations of Energy Use to Society.   The expansion of high–technology energy production has had massive effects on the organization of society (Adams, 1975; Cottrell, 1955; White, 1959), and there may be substantial social consequences of a societal choice between "hard" and "soft" energy systems for future energy development (Lovins, 1977). There is also considerable research relating energy use at the aggregate level to various measures of "quality of life" (e.g., Nader and Beckerman, 1978).

Social Processes That Contribute to Energy Policy.   Energy policy decisions are affected by political and economic interests (e.g., Engler, 1961, 1977; Rosenbaum, 1981; Wildavsky and Tenenbaum, 1981; Willrich, 1975), by ideology (e.g., Wildavsky and Tenenbaum, 1981), and by the educational, professional, and employment backgrounds of the people who have been most prominent in policy making (Nader and Milleron, 1979). These influences on policy have all been the subjects of study by political scientists and anthropologists.

The Politics of Nuclear Power.   Several social and psychological processes have crucial but not obvious roles. For example, knowledge of the empirical findings on human judgments about the acceptability of risk (e.g., Slovic, Fischhoff, and Lichtenstein, 1982) is critical for understanding why opposition to nuclear power has been so persistent and forceful despite the small number of casualties so far attributable to the technology. Also highly relevant are existing studies of the value conflicts and the social movements and organizations involved in antinuclear and pronuclear political activity (e.g., Nelkin, 1981; Rankin, 1978) and of the political processes that led to government decisions to promote a technology that was to be so vigorously opposed by a large minority of the population (e.g., Bupp, 1979).

Concentration of Control Over Energy.   Social and economic processes have led to an increasing concentration of the control over the U.S. energy system in the private sector, and, in particular, by large energy corporations. These processes are both political and economic, and they almost certainly have extensive effects upon the way energy is used in society, as well as upon the ways decisions about energy are made. This topic has recently begun to be systematically studied (e.g., Blair, 1976; Burton, 1980).

Demographic Changes in the Society.   There have been and continue to be major demographic shifts in the country that have implications for energy demand in the housing, transportation, and industrial sectors. These changes interact with each other and with other economic and social changes in complex ways. The size of the typical U.S. household is declining, the national age distribution is changing to include a larger proportion of older persons, and the migration of people and industries within the country and internationally is affecting where people live in relation to each other and to their jobs. Clearly, these social and behavioral changes affect energy use, and these relationships are beginning to be examined (e.g., Zimmerman, 1980; Abrahamse and Morrison, 1981).

Equity.   Inequities and perceived inequities have been produced by a system that allocates energy on the basis of ability to pay. Such a system enables individuals to make decisions reflecting personal tastes, but it makes energy allocation dependent on the distribution of income. The result is that some people, because they are poor, are treated in ways generally acknowledged to be unfair or unfortunate: for example, the elderly poor who are unable to buy heat in winter. Policies emphasizing further price increases for energy are frequently criticized as inequitable for these reasons. Such criticisms raise a number of empirical questions: How do people at different income levels in different regions of the country cope with rising energy prices? How effective are various policies designed to mitigate economic hardships resulting from rising energy prices? What is the potential for socially disruptive behavior as one response to perceived inequity? Information on the answers to these questions is becoming available.

To summarize, there are many important social and behavioral questions related to energy that we do not examine. But in the areas we do address, we offer new ideas for energy policy. Just as importantly, we explore a different way of looking at energy policy issues. It is our belief that adding the human dimension to the analysis of energy issues will increase understanding of how the energy system functions. Better understanding can broaden the range of policy alternatives for consideration. We believe such a broader debate can help the society organize to diminish the frequency of energy surprises, to cope better with unavoidable surprises, and to develop more effective energy policies and programs.

Notes

1. We are not the first to assert that energy has an important human dimension. The Committee on Nuclear and Alternative Energy Systems reached just this conclusion after years of primarily technical analysis (National Research Council, 1979:80):

    It is important to keep in mind that the energy problem does not arise from an overall physical scarcity of resources. . . . The problem is in effecting a socially acceptable and smooth transition. . . . Thus, energy policy involves very large social and political components that are much less well understood than the technical factors . . .

    Two of the resource groups working on that report made a preliminary effort to examine those social and political components (National Research Council, 1980; Unseld, Morrison, Sills, and Wolf, 1979), but, subsequently, there has been no serious effort to develop a fuller understanding of the human dimension.

2. Some "surprises" in the energy system, such as the timing of a power plant explosion or the assassination of a political leader that destabilizes the government of an oil exporting nation, are unforeseeable by any analytic technique. The response to such surprises also has a human dimension in that social organizations and institutions may be more or less flexible in their ability to respond to the unexpected. (The design of institutions for resiliency is discussed below, especially in Chapters 6 and 8.)

3. Information from J. Ainsworth, California Energy Commission, 1982.

4. While economists have always played a rather prominent role in energy policy making, other behavioral and social scientists have been underutilized. For example, when the U.S. Department of Energy (DOE) was created, energy research was removed from the priorities of the National Science Foundation, which traditionally funds social science research, although the DOE did not support a research effort around social issues in energy. By 1979 DOE had developed a scattering of projects addressing social issues that are related to particular programs, but it could still accurately be said (Wilbanks, 1979) that "the social sciences are the only major category of the sciences in which DOE funds

applied work but not fundamental research." The department did not seriously evaluate its capabilities in social and behavioral science at that time, and proposals for a basic social research capability within DOE have never been funded.

The department's position on social research is itself part of the human dimension of energy in that it follows naturally from the agency's institutional history. The Department of Energy evolved from the Energy Research and Development Administration and, before that, the Atomic Energy Commission, both of which had as their major purpose the development of energy technologies. DOE inherited many high-level career officials who continued to see the agency's mission as technical development, even though additional functions had been mandated to the new department. Partly as a function of that history, most of the DOE staff has training in the physical sciences or engineering, and few staff members are trained in the behavioral and social sciences. As of March 1981, after the department had been in existence for four years, only 85 of its total staff of 19,972 either had social science backgrounds or were working on issues relating to social science; only 17 of these claimed any academic training in sociology, political science, psychology, anthropology, or geography (Office of Program Coordination, 1981). It is therefore not surprising that DOE has seen little need for social or behavioral research or that its policy analysts are often surprised when the human dimension appears as a major factor in energy events. This gap in analytic capability was a major reason for creating the committee that produced this book.

5. The committee's work is not unprecedented, in that social and behavioral scientists have been examining the human dimension of certain energy issues for some time. There have been several major works on relationships between energy use and societal structure (e.g., Adams, 1975; Cottrell, 1955; White, 1959) and on the politics of energy (e.g., Blair, 1976; Burton, 1980; Engler, 1961, 1977; Lindberg, 1977; Rosenbaum, 1981; Wildavsky and Tenenbaum, 1981). There are numerous empirical studies of social and psychological issues relating to energy (for a partial overview, see Farhar, Unseld, Vories, and Crews, 1980; Nader and Beckerman, 1978; Seligman and Becker, 1981; Stern and Gardner, 1981) and some recent technical volumes of research (e.g., Baum and Singer, 1981; Beck et al., 1980; Stern et al., 1981; Warkov, 1978). These works, however, have been written either for relatively specialized audiences within the social sciences or for policy audiences concerned with particular energy issues. This book ad-

dresses a range of energy policy issues from a social-behavioral viewpoint and illustrates some ways of thinking about energy issues that have not been evident in most prominent analyses of energy policy. It may also suggest ways of thinking about other energy issues not addressed directly in this volume.

2

Thinking About Energy

The way a society thinks about energy affects the way society makes decisions about energy. In the United States, most policy analyses have failed to fully recognize the many meanings of energy in modern society. Consequently, promising policy options are often overlooked in energy debates. This chapter illustrates one fundamental way thinking about energy has been limited and shows how that limited thinking has shaped policy debates, excluding some plausible policy alternatives from serious consideration.

FOUR VIEWS OF ENERGY

Physicists have a clear definition of energy: it is a property of heat, motion, and electrical potential, and is measurable in joules, British thermal units, and their equivalents. When the concept is extended to include mass, energy can be neither produced nor consumed: its quantity is always conserved; and its quality is always declining. This concept of energy has been useful in advancing science and technology, but it is not always useful for social purposes. As a society, America does not usually think of energy in technical terms.

For most people, energy is both produced and consumed, and energy conservation is an option rather than a natural law. The popular definition of energy is not the same as the physicists', nor does is it have the same precision. When a new natural gas field is discovered or a new process is developed for extracting oil from shale, people feel that more energy is available. Even economic events can change the amount of energy meaningfully in existence, because "proved" and "probable" reserves of oil and other energy sources are defined in terms of the economic cost of recovering them from their physical surroundings. The energy of political debates— of "energy policy" and the "energy crisis"—is not the physicists' energy but rather a socially defined entity.[1]

This fact is crucial for making effective energy policy, for there is no single socially shared concept of energy. And each concept of energy has different implications for the way a society produces, controls, allocates, and uses energy. Fundamental lack of agreement about the definition of energy underlies many conflicts and policy shifts on particular energy issues. Furthermore, the dominance in most policy analysis of one view of energy—as a commodity—helps explain a pattern of conflict in recent debates about energy policy and the failure of the political system to seriously consider certain otherwise plausible energy-related alternatives. Explicit recognition of conflict over the very concept and definition of energy can be a step in clarifying issues and leading to more productive policy debates.

Energy as a Commodity

At least four quite different views of energy are widely held in U.S. society, and each contains some truth (see Table 1).[2] First, energy is often seen as a commodity or, more accurately, a collection of commodities. Energy means electricity, coal, oil, and natural gas. (To a physicist, electricity is the only energy form on this list; the others are substances that contain chemical potential energy that can be converted to thermal energy when they are burned.) When people talk about "U.S. energy supplies" or "projected energy demand," they are usually talking about this list of tradeable goods. Commodity energy consists of energy forms or energy sources that can be developed and sold to consumers. And energy is a commodity in a real sense because the society treats it that way: a significant portion of the U.S. economy has been built on trade in fuels and electricity.

The view of energy as a commodity reflects a certain set of values and beliefs; acting on this view tends to move particular interests to the center of attention. The commodity view emphasizes the value of choice for present-day consumers and producers. It assumes that such choice will allocate energy (and other commodities) effectively and efficiently. It also assumes that when prices rise, fuel substitutes will be found and that inequities that arise can be handled by ad hoc modifications to the system. It focuses analysis on the transaction between buyer and seller and away from other aspects of energy use that are external to the transaction. The interests of energy producers, along with those of consumers who have sufficient resources to participate in energy markets, take center stage. The effects of energy use on environmental values, social equity, occupational and public health, the international balance of payments, and the like are

Table 1. Four views of energy

View of energy	Important properties of energy	Central values emphasized	Interests emphasized
Commodity	Supply Demand Price	Choice for present buyers and sellers	Energy producers Consumers with sufficient resources
Ecological resource	Depletability Environmental impact Effect on other resources	Sustainability Frugality Choice for future citizens	Bystanders to market transactions Future generations
Social necessity	Availability to meet essential needs (distribution)	Equity	Poor people Poorly funded public services
Strategic material	Geopolitical location Availability of domestic substitutes	National military and economic security	U.S. energy suppliers Military

considered secondary, and people who are concerned with such effects must petition the political system for attention to those issues. Meanwhile, the interests of the participants in the transactions are advanced, since they are able to externalize the costs of those transactions.

For most of this century, the commodity view of energy has been widely accepted in the United States. It made sense to the companies that produced and distributed fuels and electricity, as well as to many of the companies' customers. The use of markets to exchange commodity energy seems to have been generally satisfactory, judging from the relative lack of political debate about it in the period from World War II until recently.

The oil embargo of 1973 reopened political debate about energy. Many people argued that a national energy policy was needed—as though the federal government was not already deeply involved. What they meant was that energy should be treated differently, as a special national priority, deserving special attention. A cabinet-level department was created to lead a visible and coordinated effort to eliminate U.S. dependence on foreign suppliers of oil. For those supporting creation of the Department of Energy, energy was no longer simply a commodity.

Energy as an Ecological Resource

A second view of energy is that it is an ecological resource. Even before the 1973 oil embargo, serious academic and policy debates about the relationship between energy use and environmental pollution indicated that energy was seen by many people as something other than a commodity. Those people argued that energy use had implications beyond the interests of buyer and seller. When energy is seen as an ecological resource, people who now breathe combustion products, as well as future generations of producers, consumers, and breathers, have a stake in how energy is managed.

The ecological resource view emphasizes certain properties of energy. Energy sources are classified as renewable or nonrenewable, exhaustible or inexhaustible, polluting or nonpolluting. Moreover, energy sources and transformations are seen in the context of biospheric systems: extraction and use have implications beyond energy—for soil, water and air quality; for climate; for the availability of other resources such as water and land; and for the health of biological communities. The view of energy as an ecological resource emphasizes some differences between energy sources that look the same from a commodity viewpoint. For example, coal deposits can be depleted, but hydroelectric resources, also used to produce electricity, cannot; and heating with oil pollutes the air and threatens to produce long-term climatic change by adding carbon dioxide to the atmosphere, but heating with passive solar technology does not.

Viewing energy as an ecological resource suggests that by using energy at the world's present rate, the present generation might be altering the environment and making the world a less healthy place for its children and grandchildren. Because high levels of energy use may irreversibly alter major environmental systems, such as climate, the resource view emphasizes frugal energy use.[3] Since the western industrial world is particularly dependent on nonrenewable energy resources (oil, coal, natural gas, and uranium), the ecological resource view puts special emphasis on careful use of those energy sources.

Like the commodity view, the view of energy as an ecological resource reflects a value system and focuses attention on particular interests. The resource view emphasizes the values of sustainability and frugality. It also values choice, but future choices have a higher priority than in the commodity view. It assumes that energy resources are finite and interdependent with other resources and emphasizes the fact that present use displaces significant costs to nonparticipants in energy transactions and to future generations. Thus, the resource view draws attention to the interests of groups that pay the costs of this energy although they are outside the market transactions—workers in unsafe mines, breathers of polluted air, energy users who pay for powerplants that need not have been built, and future generations of workers, breathers, and energy users. In this view, buyers and sellers have legitimate interests, but other groups are the center of attention. This view often leads to setting limits on market transactions through a political process: for example, by regulating or taxing resource extraction, waste, and pollution. The effort, in short, is to develop an energy system that incorporates the interests of those who are not participants in, but are affected by, buying and selling of energy.

Energy as a Social Necessity

A third major view of energy has also become increasingly important in the last ten years: energy as a social necessity. In this view, people have a right to energy for home heating, cooling, lighting, cooking, transportation, and for other essential purposes. In a biological sense, these things are no more basically essential to human life than they ever were. But society has changed greatly in the past century, and the social definition of "necessity" has changed with it. Because of technological advances and more than a generation of inexpensive and readily available energy, most people in the United States expect that they and their neighbors will have heat and lights in their homes whenever they want them. The same history has led many Americans to live more than a day's walk from work, so transportation to and from a job is needed. And many people now live in desert climates that are inhospitable without using energy to cool homes and workplaces and to import water. The changes that have increased the need for energy in the United States were not especially salient until 1974, when the long, slow decline of energy prices abruptly reversed. But salient or not, real energy needs have always existed.

Energy as a social necessity is not a new idea. Welfare programs have treated utility bills as necessary household expenses for a long time, and government utility regulators have typically sought to prevent or minimize interruptions of service. But events surrounding the 1973–1974 oil embargo made energy needs much more apparent to both consumers and policy makers. Interruptions of energy supplies threatened some areas with power blackouts, and many consumers stopped taking home heating and mobility by automobile for granted. For the poor, the rapid rise of oil prices forced choices between energy and food, clothing, or other essentials. City governments began to see the cost of energy change from an insignificant expense to one of the largest items in municipal budgets—and essential services depended on keeping fire engines, police cruisers, and other vehicles running and on keeping city buildings comfortably heated or cooled.

The central value implicit in the view of energy as a social necessity is equity. Certain energy needs must be met by society as a precondition for any further allocation of resources. This view assumes that private action will not meet these needs for everyone and that public action is essential. In a market economy, it emphasizes the interests of those who lack market power—chiefly poor people and poorly funded parts of the public sector. It also supports people who may face energy-related hardship in a crisis. In this view, the goal is to ensure a minimum energy standard for all; energy beyond what is required to meet minimum needs can then be treated as a commodity or as a resource.

Energy as Strategic Material

A fourth significant view of energy is as strategic material. In this view, the important properties of each energy source include its geographical location in the world; the political stability and orientation of the countries in which it is located; and, if an energy source is located in an unstable area, the availability of domestic or other reliable substitutes.

Energy became noticeable to the public as a strategic material in 1973 when oil was used as a political weapon against the United States. Of course, oil had always been strategic, because of the dependence of the national economy on oil-fueled internal combustion engines and on the profitable functioning of oil-related industries. But the significance of this dependence increased as the level of oil imports from the Middle East increased and with the development of international agreements to share available oil with other nations. By 1980, the strategic vulnerability of the U.S. energy system had become a central concern of federal energy policy (Lewis, 1980).

The view of energy as strategic material emphasizes a value of national security defined in terms of economic vitality and the maintenance of military strength. It assumes that energy—especially oil—is essential for an economically and militarily strong nation and that public action is required to ensure a secure supply. This view emphasizes the interests of groups that supply and demand oil for strategic purposes, particularly U.S. energy companies and the military. These interests are promoted by spending federal money for military fuel, hardware, and personnel and by protecting the overseas investments of U.S. energy companies. When, as at present, the strategic view of energy is advanced in tandem with a military buildup, the interests of the many industries that supply the military are also advanced. In this view of energy, strategic needs take priority over all others, and any debate about how to treat other demands on energy sources and federal funds must take place after those needs are met.

Conflict about Views of Energy

The most heated debates about specific energy issues are also conflicts about the nature of energy.[4] The intensity and persistence of energy conflicts reflect the fact that more is at stake than the specifics of any particular policy. Policy choices are often, implicitly, choices among different views of energy, and as such, they legitimize those views of energy most consistent with the chosen policies. Consequently, the effects of policy decisions can be more profound than the particular policies adopted. When a society implicitly accepts a particular definition of energy, the choice tends to set the terms for future political debate and define the legitimate participants; future policies are likely to reflect particular interests.

Recent disputes about government support for the commercialization of conservation and solar energy technologies are an instance of conflict between the commodity view and other views of energy. The argument against government involvement rests on the assumption that market forces will allocate resources among energy technologies more efficiently than government decisions—an assumption that is widely accepted for ordinary commodities. The argument for a government role is based on the assumption that energy is not an ordinary commodity. In this view, energy involves a special national interest—in minimizing the effects of pollution; in saving natural resources for future generations; in meeting human needs; or in reducing the nation's vulnerability to disruptions of oil imports.

A variety of government policies make sense in terms of some views of energy, but not others. Oil and gas price decontrol are appropriate policies if the goal is to allocate commodities efficiently, but such policies are counterproductive if energy is viewed as a social necessity. Energy assistance and home-weatherization programs can effectively provide for social needs, but they interfere with market allocation of energy commodities.

Standby gasoline allocation and conservation assistance for schools and hospitals provide for social necessities but interfere with markets; exhortations to save energy, fuel economy standards for automobiles, and conservation tax credits conserve resources but may disregard issues of social necessity; federally funded oil stockpiles and emergency contingency plans preserve national security but constrain markets; and deregulation of the generation of electricity would allocate that commodity more efficiently but might have significant environmental and social costs. Each of these policies institutionalizes a particular view of energy and gives other views second place.

Adoption of any policy represents a choice among views of energy. By creating or strengthening interest groups, a single policy can provide institutional and symbolic support for a whole range of policies based on a similar view of energy. In this way, energy policies help shape a society's definition of energy. The connection of policies to interests and to basic perspectives partly explains why energy debates have been more acrimonious than they would be if energy were merely a technical issue. It also helps explain why consensus about energy facts does not often resolve energy disputes.

A Shifting Foundation for Policy

Disagreement about the nature of energy makes it difficult to sustain societal effort in dealing with energy problems. The difficulty is increased because rapidly changing events continually shift the attention of the public, experts, and political interest groups from one aspect of energy to another. A glance at the history of U.S. energy policy over the past decade demonstrates these shifts in focus.

For most of this century, energy costs, resource depletion, and environmental pollution were not salient for most people. There was relatively little reason to question the view of energy as a commodity or to offer alternatives. Only since the early 1970s, when this situation changed, did spirited public debate arise. The environmental movement and the "limits-to-growth" argument focused attention on ecological resource issues, and the temporary oil shortage of 1973–1974 was seen as evidence of resource depletion. The oil shortage also focused attention on energy as a social need and on energy's strategic significance, as fuel shortages threatened the economy and many people suddenly were unable to pay market prices for energy. Thus, competing views of energy attracted attention and led to the organization and reorganization of political movements and interest groups to influence "energy politics."

In the period between 1974 and 1979, oil became more readily available and real prices dropped. The concern with energy needs and resource issues was expressed with less urgency while concern about energy as a strategic material remained acute because of the still-increasing level of oil imports. National policy makers came to see oil vulnerability as the preeminent issue in energy policy (e.g., Lewis, 1980). The 1979 oil shortfall again focused attention on the needs of individuals and municipalities because short supplies were unevenly distributed and because price increases forced hard choices on many consumers. More recent events, including a sharp decline in oil imports and a downward slide in oil prices, have again changed perceptions in the policy community. With shortages and price fluctuations receding into memory and the rapid decline of oil imports through 1981 and 1982, the arguments that the Department of Energy was not needed and that the market could handle energy became more plausible. Once again there was a shift to regarding energy as a commodity.

These shifts of perspective are likely to continue for some time. When sharp price increases for forms of energy occur, they direct attention to unmet social needs. Recurring political crises in the Middle East underline the necessity of energy and reemphasize energy's relationship to national security. And major accidents or environmental incidents associated with nuclear power production, oil refining, coal burning, or the disposal of petrochemical or radioactive wastes—which can occur at any time—remind people of the importance for ecological systems of careful use of energy resources.

Yet, despite the rapid shifts in the way experts and others perceive energy problems, the problems themselves are rather stable and persistent. Because of the dependence of Western economies on Middle Eastern oil, strategic problems will continue to surround the energy issue. Because energy prices are unlikely ever to return to their pre-1973 levels, many households and localities will continue to suffer economically.[5] Even by the most optimistic estimate the buildings sector of the economy will take decades to adjust to the price increases. And because of the time delays involved in such environmental problems as acid rain and the greenhouse effect, environmental issues will persist in energy policy, no matter what actions are taken in the immediate future.

This situation presents a dilemma for sustained policy making. Effort is necessary to deal with long-lasting energy problems, but because of changing perceptions, it is difficult to maintain such effort. Policy analysts tend to focus on only one or two aspects of energy at a time. This weakness in developing an inclusive energy policy is rooted in the failure to recognize energy as being many things simultaneously. Given rapidly changing world conditions, policies based on any one view of energy are likely to seem inappropriate when conditions change. When public officials fail to appreciate this point, they often believe that their particular views of energy can be sustained politically over time. For example, former Secretary of Energy James Edwards could defend the Reagan Administration's abrupt reversal of policy on conservation, solar, and nuclear energy by saying (quoted in Smith, 1982): "We are putting behind an era of stop-and-go policymaking." Given society's failure to recognize the complexity of energy, Edwards' reversal will not be the last. A fuller appreciation of the meanings of energy can generate a more stable policy process, appropriate to the nature of U.S. energy problems.

THE DOMINANCE OF THE COMMODITY VIEW

The four basic conceptions of energy do not have equally strong support, either in the political arena or among policy analysts. In most aspects of the national policy process, the commodity view is dominant. Dominance of a particular view of energy does not mean that it is the only view given consideration, but that other views must make special claims before being taken seriously. And in most U.S. energy policy debates, the burden of proof still remains on those who assert that energy should be treated as something other than an ordinary commodity. When these advocates succeed, they do so by winning exceptional treatment for particular situations rather than by changing the dominant perspective. Two examples illustrate this dominance of the commodity view.

The National Environmental Policy Act of 1970 mandates a procedure for evaluating the environmental and socioeconomic impacts of major federal actions. Because many energy activities come under the purview of the act, the Department of Energy conducts environmental assessments of its major programs. In practice, energy programs are usually conceived and justified on technical and economic grounds and only after they are designed and proposed are the likely environmental effects examined. If those effects are negative enough and serious enough, the program can be modified or discontinued.

Socioeconomic impacts, which include equity effects, effects on communities, and the like, are considered as part of the environmental analysis, but usually a small part. While this procedure ensures some consideration of environmental and social values, they are considered only after energy officials have developed some commitment to a program. Adverse environmental impacts are seen as requiring "mitigation" so a program can proceed; financial compensation is considered for communities disturbed by an energy project; and public antipathy toward a new project is regarded as a "barrier to implementation." In this process, the prospective project is evaluated as a given, and the role of environmental and social concerns is as a barrier. Advocates of those concerns are placed in a reactive and adversary role with respect to government officials (see Nelkin and Fallows, 1978).

The view of energy as a commodity is built into these procedures. This becomes obvious when one considers how different the policy process might be if the ecological resource or social necessity view dominated. In those cases, policies would be conceived and justified primarily on grounds of their desirable effects on environmental or on social values, within technical and budgetary limits. The primary goals would be to enhance environmental quality, to strengthen local communities, or to develop technologies that the public wants. For example, only after a policy were first suggested on such grounds would analysis be carried out to evaluate the costs of the energy services it would provide under current and projected market conditions. High cost would be considered a barrier to implementation, and the government might consider surmounting the barrier by tax subsidy, regulation, or other means.

Another example of the dominance of the commodity view can be seen in the way in which the resource conception of energy has been advocated in recent years. While the argument for careful and limited use of energy resources can be supported with evidence of the adverse environmental effects of mining low-grade fossil energy resources, burning carbonaceous fuels, disposing of radioactive wastes, and the like, the most influential recent works promoting the resource view have argued the point mainly on the grounds of economic efficiency. Thus, in discussions of energy conservation, a distinction has been developed between energy efficiency, on one side, and curtailment or sacrifice, on the other (e.g., Hayes, 1976; Yergin, 1979). A number of recent studies have emphasized that many energy users should, out of pure economic self-interest, increase their investments in energy-efficient buildings and equipment (Ross and Williams, 1981; Stobaugh and Yergin, 1979; Solar Energy Research Institute, 1981). Some writers have even attempted to redefine the commodity view to provide an argument for conservation: by defining the commodity in question as "energy services"—amenities such as heat and transportation rather than merely the fuels used to produce them—they argue that a free market for energy services would produce very low levels of growth in energy use, even in an expanding economy (Sant and Carhart with Bakke and Mulherkar, 1981). Apparently, concern with resource depletion, air pollution, and the like are less persuasive in the energy policy community than concern with economic efficiency. The advocates of energy conservation have found a way to get ecological resource issues considered, but not on their own terms.

The commodity view of energy dominates the resource and necessity views in the sense that environmental, equity, and related concerns must be asserted as reactions to policy initiatives that usually derive from the commodity view. However, the commodity view is not always dominant. A strategic consideration often dominates other concerns—as is evident in the area of petroleum allocations. In that particular debate, the burden of proof has fallen on those who hold the commodity view, arguing that the market can handle allocation in an emergency better than the government. In this sense, the strategic view seems dominant over the commodity view. Strategic considerations also dominate over the resource and necessity views. For example, there was widespread support for the expensive strategic petroleum reserve even in 1981, when energy programs concerned with environmental quality and the needs of low-income energy users were being cut severely.

While the commodity view is generally dominant, and the strategic view dominates in some policy arenas, continuing conflict demonstrates that no single view of energy has achieved universal acceptance. This is understandable since each view of energy contains some of the truth, and each has a constituency. So political debate continues, and interests that stand to benefit from each particular view of energy continue to provide support for that view while arguing for policies that promote their interests. Energy corporations and their allies support the commodity view; environmentalists and some consumer and labor groups support the resource view; representatives of people with low or fixed incomes, as well as some municipalities, support the view of energy as a necessity; and the military establishment and its suppliers support the strategic view.

A particular view of energy can be supported in several ways. A view may be promoted in the media when an advocate appears on television saying that energy efficiency can help cut air pollution (ecological resource view) or that the country might be held hostage by Arab sheiks (strategic view). A view of energy may be promoted when people with that view have access to decision makers, such as legislators or high-level officials of the federal administration. And a view of energy may also be promoted through academic research that focuses on the variables that are central to a particular view of energy. Since energy is simultaneously a commodity, a resource, a necessity, and a set of strategic materials, research can find evidence for each view. For example, research showing that energy use declined when prices rose promotes the commodity view. However, while energy use declined other things also occurred. For example, poor people sacrificed some amenities and necessities more than people in other income groups (social necessity view).[6] Each set of interests understandably supports and publicizes research that legitimizes its view of energy issues.

A significant fact of energy politics is that the supporters of different views of energy have unequal resources to promote their ideas. Proponents of the commodity view are in a much stronger position for advocacy than proponents of the resource or necessity views of energy. The most powerful proponents of the commodity view are the energy corporations—some of the largest industrial corporations in the nation. No institutions in our society have greater access to the media, to research expertise, or to decision makers. In addition, the commodity view gains the support of corporate interests outside the energy system because any challenges to this view are believed to support wider government intervention in markets, a position that is opposed by most corporate interests. And because the commodity view of energy has been dominant for so long—partly because of the resources available to its proponents—many people have never been exposed to discussions of energy that proceed from other views. Thus, the commodity view of energy is well institutionalized; it has a strong and stable fund of resources in the energy industries and an established intellectual base in economic theory and research.[7] And of course, the common experience of paying for fuels and electricity makes the commodity aspect of energy a very real part of people's lives.

Proponents of the strategic view of energy also have a strong base from which to advocate their position. Like the commodity view, the strategic view has the support of many major energy corporations, whose foreign investments are protected by governmental efforts to make imported oil supplies secure and who profit from sales of oil to government stockpiles. The suppliers of equipment to the military constitute another wealthy and influential base of support. In addition, the strategic view has a stable set of well-placed advocates in the Department of Defense of any administration. And people remember past encounters between the United States and hostile foreign governments that include the use of energy as a weapon against the United States, so they can clearly see the strategic aspect of energy.

The two other views of energy have a weaker base from which to compete. They lack politically or economically powerful supporters, and, unlike the commodity view, they do not have a well-established intellectual paradigm from which to derive policy suggestions. Relatively few people have direct experience with providing for their own energy needs or managing depletable resources, and most people do not regularly experience pollution that is visibly tied to energy; consequently, the resource and necessity aspects of energy are not as regularly or dramatically experienced as, for example, gasoline purchases.

All of the above factors have helped the commodity view of energy remain dominant in the face of challenges from the necessity and resource views. Thus, energy policy alternatives presented for public debate continue to be based primarily on a commodity view, with allowances for exceptional treatment when extreme inequity can be demonstrated or when political pressure for exceptions becomes intense.

THE NEED FOR A BROADER VIEW

The dominance of the commodity view is a problem for U.S. energy policy because it restricts vision, limiting the ability of both the public and energy analysts to explore a full range of policy alternatives. One example of this limited vision is the lack of debate in recent years over public control of energy production and distribution.

Given that the environmental and equity implications of energy decisions have been increasingly salient in the past decade, one possible development might have been a serious effort to move control over energy decisions from the oil companies and electric utilities, which allocate most of the relevant resources, to some publicly controlled body. But debate over nationalizing oil or "municipalizing" electricity has only been an occasional and minor feature of recent energy politics, even though there are many municipally owned utilities. More prominent in political debates have been attempts to exert limited public control over energy corporations: by taxing windfall profits from oil deregulation or by requiring utilities to offer energy audits and to purchase power from some small producers at favorable rates.

We do not analyze the arguments for and against public control of energy, but we do find it significant that this obvious way of bringing controversial energy decisions into the political arena has not been taken seriously in recent years. We believe that the dominance of the commodity view of energy is a partial explanation: when energy is defined as a commodity, its control is seen as properly belonging in the private sector. By contrast, if energy were seen mainly as a resource, a greater public role would be considered proper; if it were seen mainly as a social necessity, decisions about production and distribution would be seen as essentially public rather than private; and if it were seen as a strategic issue, some energy decisions would be strictly controlled by government. The issue of public control is implicit in many recent efforts to solve energy problems through collective action at the local level. These efforts are examined in Chapter 7.

There are other examples of limits on policy options that relate to the dominance of a commodity view of energy. In governmental attempts to promote residential energy conservation, various programs have been considered: tax credits, low-interest loans, several kinds of information programs, and others—all of which directly affect consumers but only indirectly affect producers. Policies that might involve governments directly in the production of energy or energy services have not often been considered. For example, the U.S. government has rarely distributed insulation materials to households and has never produced such materials. A policy of outright grants to households for energy efficiency expenditures, which has been implemented in Canada, is not considered politically feasible in the United States. As a result, such a policy has not been the subject of the serious analysis given to more complex tax and loan programs that are less likely to promote energy efficiency widely in the society. Government grants for conservation may be seen as interference in the market when energy is defined as a commodity; but if energy is seen as a social necessity, grants for insulation may be seen as appropriate government actions. Chapter 4 shows how not only grants, but also effective informational programs for energy users are foreclosed by a narrow vision in which energy is seen as a commodity, and energy users are seen as rational economic decision makers.

The dominance of the commodity view has two major effects on energy policy. First, it tends to leave the energy policy community conceptually unprepared for changes in world conditions that will, from time to time, force people to regard energy as something other than an ordinary commodity. While politicians often respond to widespread public demands that are insistently expressed, they cannot respond effectively without some appropriate policy options. A capability for analysis and policy planning that transcends immediate concerns is imperative, especially in an area such as energy, where public concerns change rapidly with world events. Such a capability must include analysis of events that may bring ecological resource, social necessity, and strategic issues to the center of public attention.

One example of unpreparedness may be seen in U.S. policies to deal with the needs of low-income households faced with rapid increases in the price of home heating fuel. Partly because this problem is often seen as a secondary effect of price increases, which policy analysts have advocated for their aggregate effects, the problems of the poor are usually left for amelioration after price increases have occurred. In the crisis atmosphere of a winter without heat, a decision to offer financial assistance to pay fuel bills is understandable—and such a decision is also quite consistent with the definition of energy as a commodity. But from the perspective of meeting energy needs, energy assistance programs are far from the best way to spend government money. Direct investments in weatherizing buildings or in improving the efficiency of furnaces are more cost-effective and can meet the needs of energy users more completely and for a longer period of time.[8]

Second, the dominance of the commodity view helps create a familiar pattern of conflict in energy policy. Technical and economic considerations are the initial bases of policy analysis, so advocates of environmental values and of poor people and poorly funded public services are most often found struggling to block policy proposals that have already been approved in government agencies. In the same conflicts, energy producers often appear as supporters—active or silent—of agency positions. A typical example is the opposition of environmental groups to nuclear power plant siting decisions: endorsement of plans by government agencies has led to lawsuits, demonstrations, and lobbying efforts (Nelkin and Fallows, 1978). Another example is opposition to utility rate increases by advocates of poor people.

Yet another example arises from treating energy resources as ordinary market commodities in the event of a major supply shortage.[9] While there may be no serious political opposition to such a policy in normal times, a severe shortage will make energy needs very prominent, and needy people and organizations will act outside the market. This might result in political action—possibly powerful enough to force a hastily drawn allocation scheme. Conflict might also move outside the political system, in widespread theft of fuel or of the money to purchase fuel. Recognition of such possibilities is part of the basis of the discussion of energy emergencies in Chapter 6.

To summarize, the dominance of the commodity view of energy tends to limit policy analysis to investigation of those social institutions and processes believed to be critical to trade in commodities. As a result, otherwise plausible policy options are often overlooked, and a characteristic pattern of political conflict is reinforced. A fuller recognition of the multiple aspects of energy in policy analysis could give society a wider range of options to consider. This wider range is especially needed in a time of rapidly changing energy conditions and public perceptions.

Notes

1. The "social construction of reality" (Berger and Luckmann, 1966) is a continuing social process about which there is fairly extensive literature. One current line of research in sociology investigates how important social concepts have come to be defined and how changing definitions produce changes in social organizations and in processes that affect people's lives. Such research has been done on the social definitions of crime (Quinney, 1970), poverty, sexual deviance, alcoholism, and a variety of other "social problems" (Spector and Kitsuse, 1977).

2. The distinction among views of energy is drawn partly from Schnaiberg (1982).

3. In this view, the goal of energy policy can be described as finding ways to accomplish the things energy does for people while putting less strain on ecological systems. This approach leads to an emphasis on providing "energy services"—mobility, space heating and cooling, mechanical work, industrial process heat, and so forth—by using less energy and especially by using less energy from depletable sources. Energy services can be provided by improving the efficiency of the technologies that, by using fuels and electricity, provide energy services. Thus, homes can be insulated, the efficiency of furnaces and motors can be improved, and au-

tomobiles can transport people the same distance using less fuel. Energy services can be provided by design, such as incorporating passive solar features in buildings or using aerodynamic design for vehicles. Even more indirectly, energy services can be provided by substituting products that take less energy to manufacture for those that use more and by substituting services for products. For those favoring the ecological resource view, these substitutes for fuel are as much a part of energy as are fuels. This is why some writers have begun to speak of conservation as a source of energy (e.g., Ross and Williams, 1981; Stobaugh and Yergin, 1979).

Minimizing strain on ecological systems implies more than efficiency in energy transformations. For many people with an ecological resource view, it also means minimizing total energy use; a choice of inexhaustible energy sources over depletable energy sources; and, among inexhaustible sources, a preference for rechanneling ongoing energy flows, for example, sunlight, over developing new inexhaustible sources with major foreseeable and negative environmental impacts, for example, the breeder reactor.

4. A related argument has been made by Robinson (1982).

5. The poorest people in the United States now spend about one-quarter of their income directly on energy for use in their homes, while the richest people spend only about 2 percent (Energy Information Administration, 1982).

6. Data gathered by the Energy Information Administration (1982) show that between 1978 and 1980, energy use in the residential sector of the economy decreased by 12.3 percent while prices rose. It is also true, however, that energy costs (price times consumption) for the poorest people in the population increased 48 percent, while energy costs increased only 17 percent for the richest people. Thus, the conclusion that energy price increases produce conservation is supported by the data; so is the conclusion that energy price increases hurt the poor.

7. The commodity view of energy is central in most current economic analyses, which proceed from neoclassical economic theory. Neoclassical economics does address many ecological and national security implications of energy, using the concept of externalities and the large body of work on the problem of public goods, and it also addresses questions of social needs for energy. However, it usually argues that such needs be met through a welfare policy rather than as part of an energy policy. The nature of such analyses demonstrates the centrality of the commodity view in neoclassical economic thinking: the noncommodity aspects of energy are treated as exceptions to an analytical rule—the functioning of

ideal markets—rather than as integral parts of a phenomenon in need of conceptualization. Just as the political dominance of the commodity view puts a burden of proof on those who assert claims based on environmental preservation or social equity, the central place of the commodity view in economic analyses puts a burden of proof on those who assert that more analytic effort should be given to understanding ecological or equity issues.

8. Weatherization is more cost-effective than energy assistance in the same sense that insulation is often cheaper than energy: since weatherization costs a building owner less over the long term than energy, it is an equally good investment for a government agency that would otherwise be paying for the energy.

9. Such a plan apparently lay behind President Reagan's March 1982 veto of congressional authority to allocate petroleum supplies in an emergency.

3

Some Barriers to Energy Efficiency

The past decade has been marked by failures to make accurate forecasts or effective policy about energy use. One of the major problems is that human thinking and action are not easy to predict. During the past several years, social and behavioral scientists have begun to understand the complex issues involved in human behavior, especially as it might affect energy use. While the data are far from voluminous, the new knowledge gives a clearer picture of the barriers to energy conservation and could make efforts to overcome some of those barriers considerably more effective. In this chapter and the next two we discuss some of these issues and the new insights. We begin with a look at the broad context of energy use in the United States.

Most observers of the energy scene are aware of how inaccurate past projections of energy demand have been. For example, there is now general agreement that energy demand in the year 2000 will be much less than almost any 1975 forecast. Projections of the need for imported oil—a critical portion of national energy use—can go out of date even more quickly and dramatically. The Department of Energy estimated in late 1980 (Lewis, 1980) that continued federal support of energy conservation would help reduce oil imports from the 1979 average of 7.9 million barrels per day to an average of 6.7 million per day by 1990 and—with optimistic estimates of domestic supply—that imports might fall to 3.8 million barrels per day by 1990. But less than two years later, in spring 1982, the United States was importing only 3.5 million barrels of oil per day (Martin, 1982).

Many factors combined to embarrass the experts. The national economy failed to grow as expected for a number of reasons only partly related to energy. Demand was held down by high oil prices that rose faster and stayed higher than had been predicted after the first round of price increases in 1973–1974. In the 1970s, light manufacturing and service industries grew while heavy manufacturing declined, resulting in decreased energy demand per unit of economic production. In addition, the technological trend toward more energy-efficient equipment continued, and increases in public awareness of energy and in energy-related government programs contributed to changes in demand.

The relative importance of all these factors is not clearly understood and is still a matter of debate. Figure 1 shows an analysis of recent changes in U.S. energy demand. While the estimates of the effects of particular influences on demand would vary under different analytic assumptions, what is striking is the overall difference between estimated demand and actual demand, a decrease of 27 percent.

Much recent evidence, then, leads to questions of why energy demand is so low. But the issue is still more complex, because other evidence raises questions of why energy demand remains so high. Specifically, several studies reveal that substantial investments have not been made that would, by substituting technology for energy, lower overall costs for energy users (e.g., Office of Technology Assessment, 1982; Ross and Williams, 1981; Sant, 1979; Solar Energy Research Institute, 1981; Stobaugh and Yergin, 1979). There is also reason to believe that even if present market conditions and levels of government involvement persist for many years, much of this investment will not be made. For example, a panel of experts convened in 1981 at the National Academy of Sciences estimated that only between 30 and 80 percent of economically justified investment would be induced by energy price signals (National Academy of Sciences, 1981). Similarly, a detailed study of city buildings by the Office of Technology Assessment (1982) concluded that by the year 2000 only 38 percent of the energy savings achievable by investments that are economically justified under present conditions will occur in that sector.[1]

There is much left to learn about the behavior of energy users in the United States. The evidence of this imperfect understanding comes at a critical time for U.S. energy policy because recent changes base policy even more on assumptions about the behavior of energy producers and users that are increasingly being questioned. In particular, current policy assumes that the profit motive will encourage producers to develop and market technologies that will save users money at current energy prices and that economic motives will also spur energy users to purchase and use those technologies. This belief in the market persists in spite of the evidence that institutional barriers to investment in energy efficiency do not yield to clear market signals (e.g., Bleviss, 1980; Blumstein, Kreig, Schipper, and York, 1980; Office of Technology Assessment, 1982; Schip-

Energy Trends in the U.S. Economy

- **ENERGY USE** in 1981 was down **28 quads** from the base case.

- These **ENERGY SAVINGS (27%)** are the result of several factors:

  - **SLOWER ECONOMIC GROWTH (11 quads)** .... GNP slowed from **4.0** to **2.7** percent per year after 1972;

  - **HIGHER ENERGY PRICES (8 quads)** .... motivating accelerated efficiency improvements, better energy management, and changes in consumer preferences;

  - **TIME FACTORS (3 quads)** .. improved energy efficiencies consistent with trends of the last several decades;

  - **OTHER FACTORS (6 quads)** .. accelerated efficiency improvements not directly related to increased energy prices, including:

    - **accelerated economic trends** away from heavy industry toward the service and information industries, and lighter manufacturing;

    - **technological advances:** R & D, turnover of capital stock, expansion of modern businesses, and other recent innovations;

      - **energy shortages:** gasoline lines, fuel allocations and natural gas curtailments;

      - **deregulation** of energy using industries — airlines, trucks, and railroads — allowing more flexibility to become energy efficient;

      - **increasing public awareness,** through the news and other media, of the energy crisis and solutions;

      - **conservation programs** of utilities, the private sector, and state and federal governments, including tax credits and deductions for conservation investments;

- **FEDERAL PROGRAMS** and **TAX CREDITS** probably account for less than **5** percent of the overall increase in energy efficiency per unit of GNP due to **ENERGY PRICE INCREASES, TIME, and OTHER FACTORS.**

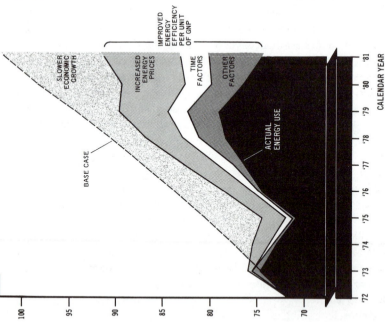

Fig. 1. Energy Trends in the U.S. Economy

SOURCE: Office of Policy, Planning, and Analysis (1982)

per, 1976) and that energy users, even within their range of choice, are often less than fully responsive to price signals or to information about how they can cut energy costs.

ENERGY INVISIBILITY AND ITS LEGACY

Because of the way energy-using technologies work, it is often difficult for energy users to take effective action when rising prices provide a strong economic motive to save energy. This is in large part because the history of material progress over the last century has made energy sources less costly in real terms and has made energy flows invisible to energy users. This history has had two major consequences, which can be called energy unawareness and energy invisibility.

We can illustrate this history with the example of home heating. In the nineteenth century most homes were heated by wood stoves. Frequently the person heating the home was also the person who chopped and stacked the wood, and the wood also had to be retrieved from stacks and loaded into the stove at fairly frequent intervals. The energy required to heat the home was quite visible to the homeowner, and the effort required to chop and stack the wood and load the stove was roughly proportional to the energy used to heat the home.

Next came the coal-fired furnace. Coal was delivered by a company rather than obtained by a homeowner. However, the coal still needed to be shoveled into the furnace, so personal effort was still required. And the daily ritual of feeding the coal furnace made the householder aware of the decreasing stack of coal; the energy flow was still quite visible.

Next came the development of the oil-fired furnace, which was marketed on the basis of its increased convenience over the coal-fired stove. The oil furnace did not require the daily tending that the coal furnace did: the component of personal effort was removed. And the flow of oil from the storage tank into the furnace was invisible to the homeowner; all that was visible was the occasional arrival of the oil truck to deliver the oil.

Next came the furnace fed with natural gas. Here the inconvenience of delivery was removed: fuel transfer was entirely effortless and entirely invisible.

Finally, electrically heated homes were introduced. Like gas, electricity offered effortless, invisible heat. In fact, this effortless invisibility has been a major source of the popularity of electricity as an energy source. And not surprisingly, it formed a good part of the successful marketing effort for electricity. Per capita consumption of electricity in the United States doubled every ten years for decades. Today, the only visible aspect of energy for most households is the bill.

This long trend has had various effects. It has allowed people to be isolated from discomforts in the physical environment and to increase their sense of well-being. People commute from thermostatically controlled heated or cooled homes to heated or cooled workplaces in heated or cooled private cars. At home, people use picture windows to let in the visual aspect of the outside world without letting in any of its nuisance properties. This style of life implies the consumption of great amounts of energy, but when prices were low, there was no pressing need to be concerned about energy costs. The low economic cost and easy availability of energy made energy users relatively unaware of energy. As a result, energy was not a salient feature in family decisions about purchasing homes and automobiles or in organizational decisions about designing buildings or maintaining equipment—decisions with major implications for energy use.

This sort of energy unawareness can be reversed, given time, if price increases or other stimuli are strong enough to make energy salient. But the consequences of low-cost energy and material progress go deeper than unawareness. Energy has became invisible to consumers, so that even with some heightened awareness, they may be unable to take effective action. This is what we mean by the legacy of energy invisibility.

Some Roots of Misinformation

Consider the fact that for most households the only visible thing about energy is the bill. In some ways, there is less useful information in an energy bill than in the actual flow of fuel into a furnace. Energy bills are received relatively infrequently, and they generally aggregate a variety of uses into a single number. To illustrate the effect of this, Kempton and Montgomery (1982:817) ask people to imagine the parallel situation for grocery bills in

> a store without prices on individual items, which presented only one total bill at the cash register. In such a store, the shopper would have to estimate item price by weight or packaging, by experimenting with different purchasing patterns, or by using consumer bulletins based on average purchases.

Obviously, few shoppers in such a store would be well informed about which changes in their purchases would most effectively lower grocery bills without sacrificing their essential items.

Yet for energy bills, especially electricity bills, that is the situation. A single bill combines the charges for several appliances, for lighting, and possibly for water heating, space heating, and cooling. It is impossible to tell, without careful monitoring and experimentation, how much of the bill results from each use or how much the bill could be decreased by using any particular appliance less or replacing it with another model.

The evidence shows that many energy users are ill-informed under this billing system. For example, most people cannot correctly rank the energy consumption of various household appliances and heating and cooling systems (Becker, Seligman, and Darley, 1979; Kempton, Harris, Keith, and Weihl, 1982; Mettler-Meibom and Wichmann, 1982). Some of the inaccuracies are explicable in terms of energy visibility: when energy use is visible, people think relatively more energy is being used. For example, people tend to overestimate the energy consumed by household lighting, which is literally visible, and their awareness is reinforced when they turn lights on or off. They tend to underestimate the much larger amount of energy used by water heaters—whose energy consumption occurs out of sight and takes place without human intervention. A survey of 400 families in Michigan found that the average householder believed erroneously he or she could save twice as much money by reducing lighting as by using less hot water (Kempton et al., 1982).[2] In addition, householders are generally unaware of technical options for saving energy, especially by modifications in the house heating plant. This pattern, which is understandable in terms of energy visibility, was also observed in a study in West Germany (Mettler-Meibom and Wichmann, 1982). The trend toward automating energy use for space heating and cooling and other purposes has bought convenience and freedom from an unpleasant environment at the cost of knowledge of energy systems. The price was small when energy was cheap, but now that consumers have a strong economic motive to cut energy use, they do not have the vital knowledge needed to do so. Not only do people make systematically wrong estimates about the energy use of various appliances, but when they act on this misinformation, (by turning off lights, for example) and find their actions have little or no discernable effect on energy bills, they tend to reject the whole idea of energy saving.

Energy invisibility also makes it difficult for an energy user who wants to save energy to learn what to do by trial and error. In a wood stove, the fire burns longer when the logs are placed properly; this can be demonstrated with only a few armloads of wood. In a modern gas or oil furnace, however, it is much harder to see the results of attempts to save energy: the effects of adjusting a burner or replacing a filter are directly observable only by a trained technician with the proper equipment. Energy saved by adjusting a thermostat or adding insulation is only observable by careful monitoring of energy use over time, and it is not usually possible for the user to assess the saving without great effort.

The energy bills that supply most information about energy use do not give high-quality information. Gas and electric utilities bill monthly (or bimonthly) for heating fuel, and fuel oil suppliers usually bill irregularly, with each delivery. These billing systems tie energy use to payments, so householders typically measure their use of electricity and gas in "dollars per month" rather than the more technically useful "kilowatt-hours (or therms) per degree-day" (Kempton and Montgomery, 1982). It is likely that many organizational managers also use such a budget-based unit of measurement. Insulation, thermostat adjustments, and furnace overhauls may reliably save therms per degree-day, but such savings may not be clearly reflected in dollars saved per month. A building owner who insulates in October may see rising heating bills for months and not be able to tell if the investment was worthwhile. Even if the owner is careful enough to make year-to-year comparisons, changes in weather and fuel prices will almost certainly confuse the message. When monthly bills are kept at the same level year-round through the use of a budget payment plan, the effects of energy-saving actions are even harder to observe and evaluate.

Energy invisibility does not make it impossible to monitor the effects of attempts to save energy in buildings, but it does make it very difficult. If high prices or personal interest provide sufficient motivation, it is possible to look beyond the bottom line of the energy bill and combine the energy-use information from a bill to weather information from a local newspaper to get a useful index of energy use. But only a very small percentage of energy users will make such efforts. The majority, who do not take this trouble, are disappointed with the results of their attempts to save energy (Kempton and Montgomery, 1982). This disappointment frequently leads to discouragement and a feeling of helplessness that makes future action less likely.

Structural Effects of Invisibility

Another part of the legacy of energy invisibility is a diminished technical capacity of energy users to respond effectively to the stimuli of price and shortage. Freedom from concern about energy has produced structural changes in energy-consuming equipment. Central heating and cooling systems, for example, allow people to move freely from one room to another without thinking about energy. But now when consumers are motivated to save energy, few people have the option of saving fuel by closing off unused rooms while retaining comfort in a smaller space. When it became possible to achieve effortless control of the internal environment, architects began to design apartment and office buildings with windows that cannot be opened. Thus, residents and workers cannot save energy in these buildings by using natural ventilation; the need for air conditioning was literally built in. Because of such changes in the national stock of energy-using equipment, no amount of energy awareness can quickly or easily reverse the effects of years of energy invisibility.

Finally, energy invisibility stands in the way of decisions to invest in energy efficiency because "seeing is believing." The design and construction features that make buildings, automobiles, appliances, and industrial equipment more energy-efficient also tend to be invisible. Insulation in walls, flame-retention heads on oil burners, aluminum in automobile bodies, and extra windings on electric motors—all save energy without being visible. But because people can't see them, they are less likely to believe they save energy. Building contractors report that it is easier to sell a new home with visible solar collectors on the roof than one with passive solar design, added insulation, or other less visible energy-conservation features, even though yet those conservation features are generally more cost-effective than active solar equipment.

The problem of energy invisibility is greater with some fuels or uses than others: the effects of invisibility are greater in buildings than in transportation. While most energy users think of gas and electricity in dollar units, gasoline is typically understood in gallons (Kempton and Montgomery, 1982). This is a more useful unit for understanding and modifying energy use. In addition, for many energy users, gasoline is purchased more frequently than other fuels, and the payment is visibly connected to the act of consuming energy. With the advent of self-service gasoline stations, more drivers are even pumping their own gas. Because gasoline use is probably more visible than most energy uses in buildings and, possibly, in industrial production, gasoline consumption is more responsive to price signals than energy for residential consumption.

In summary, the legacy of energy visibility makes it difficult for energy users to act effectively, even when rising prices make them keenly aware of energy. How much difference does this make? Winkler and Winett (1982) reviewed nineteen sets of data from experimental studies in which households were informed frequently (daily, in most of the studies) about how much energy they were using. Such feedback makes energy more visible in that it allows people to modify their habits and quickly see what changes will effectively cut their bills. Feedback led to savings of up to 20 percent, compared with similar households adapting to the same energy prices without the feedback on their use of energy. Furthermore, the effect of feedback was greatest when the cost of energy was highest (see Figure 2). When costs were calculated as a percentage of income, the relationship was even stronger. Thus, although rising costs produce energy savings, they produce much greater savings when energy users can see the effects of their actions. To put it another way, the cost of energy invisibility rises even faster than the cost of energy. The policy implications of energy invisibility are discussed in Chapters 4 and 5.

PROBLEMS OF ENERGY INFORMATION

It is important for energy users to have accurate information as a basis for action, and such information is available from many sources. But energy users are skeptical about most of what they see and hear. And no matter how carefully data are collected and estimates are double-checked, information alone generally will be insufficient to get energy users to change their stock of energy-using equipment or to use it differently, in ways that will save them money.

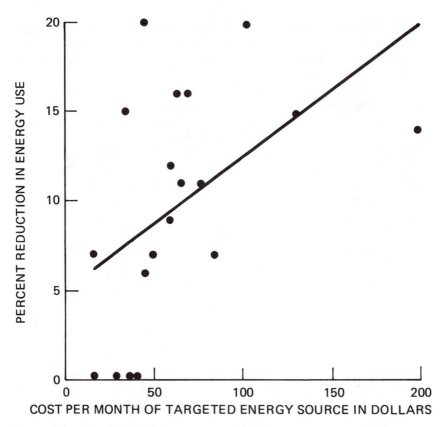

Fig. 2.   Effectiveness of energy use feedback in reducing consumption as a function of household cost for targeted energy source
SOURCE: Data sets analyzed by Winkler and Winett (1982)

A central problem with information for energy users is uncertainty: accurate estimation of the net costs of energy options depends on future economic conditions in general and, in particular, on future energy prices and availability. Both of these variables have recently been difficult to predict. Thus, large investments in energy-using equipment are seen as speculative. Their payoff depends on future energy prices in general, future prices of specific fuels relative to each other, future inflation rates, equipment costs, and changes in governmental incentive programs.

The best an energy user can hope for is to reduce uncertainty to a calculated risk. But householders and small business operators rarely have the time and money to estimate all the important variables; it is too difficult and time-consuming to search out and evaluate the available information. Consequently, they are likely to avoid making costly errors of commission by continuing past practices. People may also make decisions on the basis of the closest, rather than the best available, information. For example, a homeowner might act on the passing comment of a neighbor to the effect that even with new attic insulation, energy bills did not go down, or on the experience of warmth and light in a friend's solar greenhouse, or on the good feeling he or she gets from the idea of having solar collectors on the roof, or on the sales pitch of a contractor. In an uncertain environment, decision makers often take shortcuts in gathering information. As a consequence, they have usually made fewer investments in energy efficiency than would be justified if they had and used accurate knowledge about the best available equipment. And the investments that they do make may not be those most likely to save money.

The irony is that with the present level of uncertainty, energy users need information more than ever: about whether to be concerned about shortages, about whether or how to prepare for price increases; and about the initial and operating costs of available energy-using equipment. An alternative to detailed calculations is to receive credible information— information about the future from an expert, reliable source that can be believed. For several reasons, however, information is not credible for most consumers, no matter how well documented it is.

Diversity of Consumers

Energy information is often discredited because it is not appropriate for many of the people who receive it. This is almost unavoidable because the needs of energy users vary so widely: what saves energy for some users is wasted effort for others. Automobile and truck drivers vary in the ways they use their vehicles, their needs for cargo space, and the kinds of traffic they confront, so the same advice will have different effects for different drivers. Commercial buildings vary greatly in their uses, hours of operation, and methods of management—so general advice is probably useless. And there is also a tremendous variation among household energy users.

In homes, different fuels are used for different purposes, and the price of the same fuel may vary widely from one area to another. Energy needs vary greatly with climate, the size and manner of construction of the home, and the ages and employment status of the occupants. Local regulations or policies may promote a particular energy-saving practice (such as passive solar design) in some areas while impeding it elsewhere. Households have different appliances and different habits of using them. Some households pay directly for all the energy they use, some for none of it. Some have direct control over their heating and cooling systems, and others, chiefly in multifamily dwellings, do not. And some homes come close to their blueprint specifications while others, because of age, disrepair, or faulty construction, do not.

In the language of marketing, the residential energy market is "highly segmented." With a great deal of variability, most simple rules of thumb, such as "the first thing to do is weatherstrip," will be wrong for some households. Other advice, such as to turn down thermostats on furnaces and water heaters, although generally applicable, will be irrelevant for those who lack control over the equipment. A longer list of supposedly informative statements might include some information that is wrong for almost every household.

Because of the market segmentation, people who accept general energy advice, from whatever source, risk disappointment. The disappointment tends to discredit the source of the information. There has been an attempt to circumvent the segmentation problem by offering energy audits of private homes and other buildings in order to make recommendations appropriate for the climate and the special features of the building. While such information is almost certain to be more accurate than general advice, even it is attended with uncertainty. There is evidence, for example, that actual buildings do not use energy the way simple physical models predict (Beyea, Dutt, and Woteki, 1978). There is also the evidence that even in identical buildings some families of the same size use twice the energy others use, as a function of behavioral differences that are only partly understood (e.g., Sonderegger, 1978). Thus, even sophisticated and individualized recommendations by trained auditors or other energy experts will be wrong some of the time. Energy users who know this will be appropriately wary of information; those who do not will sometimes be satisfied, but other times they will become disillusioned. The disillusionment, when transmitted to neighbors, family, and friends, increases the general skepticism.

Conflicting Information and Policies

Despite the limitations of information, expert advice can help energy users who are concerned with minimizing energy costs. And many formal information sources exist: various governmental agencies at federal, state, and local levels offer conservation information, as do many electric and gas utilities, heating oil suppliers, heating and insulation contractors, and representatives of the petroleum, automobile, and building industries. Private energy service companies that claim to offer expert information are also available in some communities. Unfortunately, these sources offer conflicting expertise, and it is inevitable that they will continue to do so. Since there can be no incontrovertible source of knowledge in a fluid situation, there is room for disagreement—and incentive for disagreement continues to be provided by conflicting economic interests and divergent bureaucratic and disciplinary points of view.

In the Northeast, for example, heating oil suppliers emphasize the savings from improving furnace efficiency, while gas suppliers have stressed lower fuel costs in an attempt to get people to switch from oil heat to gas heat. Neither supplier, however, has much interest in showing how much energy a building owner can save by insulating or weatherizing. In these cases the economic interests are evident, so a wise consumer will be suspicious.

But there are also conflict and contradiction among supposedly disinterested government sources of expertise—also for understandable reasons. Agency personnel want to protect their programs' missions. When the Department of Energy mounted a conservation effort, advocates of the electric car competed for funds with proponents of car pooling and other energy-saving alternatives: consumers were being told to change travel behavior by one part of a government agency while being told by another part of the same agency that new technologies would soon make such changes unnecessary. More recently, statements from DOE have assured Americans that market forces will act automatically to solve the country's energy problems, but this position discredits all the department's previous statements, as well as many of its active programs and policies in the area of conservation.

The budget process encourages governmental agencies to oversell their programs. This practice often leads to the programs' being discredited in the future. Proponents of new energy production or conservation programs tend to sound alarms: "We are running out of energy," "The balance of payments is crippling the nation," "Dependence on Middle Eastern oil threatens national security." Whether true or not, such statements tend to strain the credulity of people who have not experienced oil shortages or realistic threats of war. Proponents of new energy programs also tend to overstate the benefits of their programs, leaving the impression that the Alaska pipeline, or synthetic fuels, or price increases, or some other policy choice will "solve" the energy crisis. Such rhetoric suggests that the new program is a panacea. It undermines exhortations by other agencies—or other parts of the same agency—to conserve energy, and it sets the stage for disappointment and discrediting of the agency when the program falls short of unrealistic expectations.

The same sort of overselling has also happened with government conservation programs. For example, government sources once recommended cellulose insulation because of its high resistance to the passage of heat. Then, after several homes were insulated, cellulose was discovered to be a fire hazard. Apparently the recommendation had been made without sufficient testing to ensure safety and earn consumer confidence. Such tragic blunders tend to reduce the credibility of all government agencies that promote energy conservation.

The credibility of government information is also compromised by pressures from political and interest groups. Such pressures have influenced agencies' positions, sometimes even leading to shifts on points of fact. For example, when the initial regulations for the Residential Conservation Service (RCS) were being drawn in 1979, the decision was made to recommend all "cost-effective" energy measures but one—householders were not to be advised to switch to less expensive fuels. That information had proved to be too politically sensitive.[3] When the Reagan Administration came to power in 1981, pressure for deregulation from the utility industry led to new regulations that further reduced the list of recommended conservation actions.[4] Similarly, the official position on whether rail transportation is, in fact, more energy efficient or less energy efficient than automobile transportation changed when the administration changed.[5]

We do not mean to suggest that the federal government has no credible energy information to offer. On the contrary, several federal agencies are highly credible sources of certain kinds of energy information. The Energy Information Administration is the best source of information on current patterns of energy use in both residential and commercial buildings, and the Census Bureau collects credible information on fuels used in residences and the distances people travel to work. Such information is obviously essential for major policy decisions affecting energy use.

The conflicts in government go beyond information; they are embodied in policy—and news about policy becomes part of the information available to energy users. In transportation, for example, the federal government has set standards for the fuel efficiency of new cars, yet it offered financial support for Chrysler Corporation, a company that was in trouble partly because of its failure to produce fuel-efficient cars. The government also fought hard for the agreement to limit imports of relatively fuel-efficient cars from Japan. Similarly, Amtrak advertises the fuel efficiency of rail transport, yet the government has cut back its rail subsidies, and supports highway programs. At the municipal level, budget constraints limit support of energy-efficient public transit systems.

We are not concluding that any of these policies were wrong or misguided; conflicting priorities are inevitable in a complex government. But it is important to be aware that such decisions produce profound, if unintended, effects on perceptions, attitudes, and behavior of energy users by implying that government officials do not take energy efficiency very seriously—beyond making public pronouncements.

How are energy users to know which agency experts to believe or which policy best promotes their own interests or local or national needs? Could the problem of inconsistency be solved by forcing government officials to promote a common view? We think not. Even if government could be internally consistent, private parties could and would dissent. There is adequate expert opinion available on different sides of the energy issue to confuse most energy users. And with the powerful interests that are involved in the energy debate, the conflicting opinions will be broadcast: if government does not discredit itself, others will. It is safe to assume that individuals and organizations will continue to receive conflicting information on how to get their energy services most economically. Energy users will have to cope with this situation by choosing one of the conflicting sources; by using political pressure to obtain information they can trust; by creating institutions that will provide the information they want; by rejecting all the expert advice and relying on nonexperts; or by doing nothing.

Trust in Information Sources

We have pointed out that since energy users cannot get *accurate* information about the ultimate comparative cost of different energy options, they will rely on the most *credible* available information. In fact, there is a large body of well-controlled experimental literature showing that the effectiveness of a given message depends on the credibility of the source of the message (Hovland, Janis, and Kelley, 1953; McGuire, 1969, 1983). Credibility involves a combination of expertise in the content of the message and trustworthiness. Other things being equal, the greater the expertise and trustworthiness of the communicator, the greater the impact on the audience. A given message, when attributed to a person of high credibility, produces greater attitude change in the target audience than the identical message when attributed to a person who is generally regarded as either inexpert or untrustworthy (Hovland and Weiss, 1951; Aronson and Golden, 1962; Aronson, Turner, and Carlsmith, 1963).

Many of the sources offering energy information to households are considered by the public to be expert, so in this sense, they are equally likely to be effective. When the experts disagree, however, users are most likely to rely on the sources they trust. Trust in sources of energy information does seem to make a difference.

There is some anecdotal evidence on the trust issue drawn from the experiences of community-based programs that offer energy conservation services for households. In low-income communities, both in cities and rural areas, grass-roots energy groups have gained the trust of residents because of their personal contact with the community, where more formal institutions might well have been ignored (Stern et al., 1981). But more convincing evidence comes from experimental research. In one study that has experimentally investigated the source of energy information, Craig and McCann (1978) sent a pamphlet describing how to save energy in home air conditioning to 1,000 households in metropolitan New York. Half the households received the information in a mailing from the local electric utility, the other half in a mailing from the state regulatory agency for utilities. The following month, households that had received pamphlets from the regulatory agency used about 8 percent less electricity than households that had received the identical pamphlets from the local electric utility company. Since air conditioning accounts for only part of each household's use of electricity, the 8 percent savings is clearly an underestimate of the effect of the information source on behavior. And since utilities are often perceived as particularly untrustworthy (see, e.g., Brunner and Vivian, 1979; Milstein, 1978), the trustworthiness factor has potential policy implications for energy information programs. Craig and McCann's (1978) findings also suggest that some organizations may be unable to influence some consumers, no matter how expert they are and no matter how accurate their information. This finding should not be taken to suggest that all utilities will always be ineffective in delivering information; trust in utilities varies from region to region. But it is a warning against placing too much reliance on expert information alone. People respond not only to information, but also to their perceptions and evaluations of the source of the information.

Are there any generally trustworthy sources of information? Levels of trust in sources of energy information are not uniform. Different subgroups have been affected differently by their experiences with particular institutions. Some homeowners have well-founded impressions of building contractors, for example, and people on welfare have some basis for judging how much to trust an energy program administered by the local welfare agency. Consumers' experiences with conflicting information about energy and with receiving irrelevant information, such as detailed information on home insulation that is offered to apartment renters, also affect the willingness of different groups to trust information from particular sources. For these reasons it is a mistake to believe that some agency or organization can be found that could, given good information, effectively inform the public.

THE SYMBOLIC MEANINGS OF ENERGY USE

Public debate on energy issues is couched in symbolic language: "the moral equivalent of war," "freezing in the dark," "energy independence," and so forth. Much of this symbolic debate seems to associate energy with control, power, and freedom. In the traditional view, these values are associated with ever-increasing energy supplies and consumption: energy confers goods and services, upward social mobility, and "the good life". Shortages of energy are seen to portend national weakness, economic stagnation, and an end to "the good life." The rhetoric of energy independence was used to argue for energy production, and the "moral equivalent of war" was a national mobilization of capital to produce energy and of consumers to sacrifice for a common good. Rhetoric about getting government "off the backs" of individuals and private enterprise suggests that industry, freed of certain regulations and tax burdens, will produce more energy and that more energy will mean national strength and individual prosperity.

There is another view of the relationship between energy use, freedom, and control. In this view, large-scale energy development does not enhance personal or familial control; rather, it shifts control from individuals, families, and neighborhoods to large-scale corporate organizations and to state, federal, and foreign governments. Efficiency in energy use and the development of locally available renewable energy sources are seen as essential to retain or regain control. Shortages of supply are the proof that energy means vulnerability and dependence; "freedom" is freedom from control by distant suppliers of fuel and electricity. The rhetoric of self-help and local self-reliance has been used to motivate local energy conservation efforts in cities such as Fitchburg, Massachusetts and St. Paul, Minnesota. It also figures prominently in the title of a recent study that argues for the value of energy efficiency: *Our Energy: Regaining Control* (Ross and Williams, 1981).

Freedom and control are powerful psychological symbols. Accordingly, the battle over symbols—the association of control and freedom with energy use or with energy saving—is almost certain to have tangible implications for the behavior of individual energy users and for the public acceptability of alternative energy policies. Experimental research in laboratory and field settings has shown that when external demands or regulations put pressure on people against making a particular choice, people resist this threat to their freedom—by increasing their preference for other options (Brehm and Brehm, 1981; Mazis, 1975) and even by increasing the behavior the pressure was intended to prevent (Reich and Robertson, 1979). So, all other things being equal, it is predictable that public response will be negative when a program to modify energy use is seen as a threat to freedom. The loud public outcry when President Carter proposed to raise gasoline taxes if the nation failed to meet conservation targets was one good example of the results of threats. By contrast, the public readily accepted a five-cents-a-gallon gasoline tax proposed in 1982, when the tax was not presented as a punishment for failing to conserve energy but as a necessity. Another example was the 1975 army experiment with a device that created physical resistance when drivers tried to accelerate cars or trucks too rapidly. The device was soundly rejected by drivers, and in about 10 percent of the cases the drivers simply disconnected the gadget (Thomas, Petter, Spurway, and Etzler, 1975). Possibly as a result, there was no net energy saving, and the device was quickly abandoned.

There is also the positive side of freedom and control. Just as forced restrictions tend to cause "psychological reactance," allowing people to make choices—even in simple situations—can have quite powerful effects on their well-being (e.g., Langer and Rodin, 1976; Rodin and Langer, 1977). Researchers at the Center for Energy and Environmental Studies at Princeton University have applied this principle to energy conservation (Becker, Seligman, and Darley, 1979). The researchers were studying people's resistance to installing automatic day-night thermostats. Reasoning that the resistance was due to people's not having enough control over the temperature settings, the researchers had the thermostat redesigned so that residents could temporarily override the system. That simple modification made the automatic thermostats much more attractive to users: it gave residents control by enabling them to adjust the system whenever they deemed it essential.

Such evidence suggests that public response to new energy policies and technologies may be greatly influenced by the way innovation is related to freedom and choice. The symbolic meaning of energy innovations depends partly on the intent of policy makers, but it also depends on the state of the energy system at the time a new technology or policy appears. On one hand, conservation measures that occur in response to shortages and within tight timetables are likely to be seen as coercive. They underscore the linkage of conservation with loss of control and are likely to meet with resistance. On the other hand, conservation in response to the relatively slow pressure of rising prices may lead most consumers to associate conservation with increased control.[6] Policies that facilitate conservation through efficiency improvements can be expected to reinforce the linkage of conservation and control, while policies that make it more difficult to get money to invest in energy efficiency are likely to lead to an association of conservation with dependence and loss of freedom. Thus, the symbolic meaning of energy consumption and conservation affects public reactions to energy policy—and policy, in turn, has some influence over what the symbolic meanings are.

LIMITED CHOICE

Most of this discussion implicitly assumes that consumers are basically free to choose among actions that imply different amounts of energy use. However, as many writers on the subject have noted (e.g., Schipper, 1976), consumers' choices are limited in several important ways. We alluded to some of these limitations in the discussion of energy invisibility; here, we briefly describe some of the commonly recognized limitations.

The Roles of Intermediaries

It is well known that consumer choice about energy use is limited in many ways by the various actors that we refer to as intermediaries (see Chapter 5). Intermediaries are individuals or organizations that effectively make choices for energy users but whose interests may not coincide with those of the people or organizations that pay for energy. Intermediaries limit choice when they foreclose options. Some intermediaries make actual purchases for the ultimate consumers, as when builders or building owners select heating and cooling equipment and built-in appliances for new homes or rental units. Other intermediaries limit choice indirectly. Owners of existing rental buildings, for example, make it difficult for occupants who pay for heat to lower their heating bills because the occupants are unlikely to make investments to improve the operating efficiency of someone else's property. Household energy use is also constrained by the past actions of local governments, which decide what transit services to provide and where to allow stores and work places to be built. Standard-setting organizations affect energy use by others when they set requirements for manufacturing products such as heaters, and for constructing buildings.

Manufacturers of Consumer Products

Energy users have limited control over the assortment of energy-using products from which they can choose. While lack of consumer interest can guarantee the failure of a product, consumer interest cannot guarantee that a new product will be manufactured. The producers have the initiative, and producers do not necessarily profit by marketing energy-saving products. In the home appliance market, for example, highly energy-efficient products are usually more costly to produce because of the extra materials needed to provide insulation and to make more energy-efficient motors. Some people would purchase such appliances despite the added initial cost. But without expensive advertising to educate energy users, a manufacturer who chooses to produce an efficient appliance might well lose the market to the manufacturer of an inefficient machine with a lower price tag.[7] Because consumer preferences for energy efficiency cannot reliably be expected to induce manufacturers to produce efficient equipment, government programs have recently required energy efficiency labels to be attached to major household appliances. The rationale for this approach is that a mandated program to educate purchasers about the lifetime costs of owning and operating appliances will make consumer preferences more effective in the market.

But even with some knowledge about the energy cost of a product, consumer preferences for efficient equipment are not easily translated into the manufacture of such products. Producers may seek less expensive alternatives to filling consumer desires. In 1979, when motorists became concerned about gasoline shortages and began to demand more fuel-efficient vehicles, one response by U. S. manufacturers was to advertise large cars by emphasizing their ability to get many miles per tankful; of course, these ads downplayed the size of the tank. Another response was to hold down the prices of large models so that advertising could truthfully show that even though a large car costs more to run, it is noticeably cheaper over its useful life than some small, more fuel-efficient, but higher-priced models.

Congress recognized the importance of automobile manufacturers' influence on energy use when it established fuel economy standards for the industry. At least one prominent study (Hirst, 1976) concluded that this was the single most effective available policy for conserving energy in the transportation sector. It took this government pressure of legislated standards combined with regulations requiring the posting of fuel economy data on new cars, fierce foreign competition, and intense and persistent consumer nonresponse to available U. S. models to prod the U. S. auto industry to produce fuel-efficient models.

Long-Lived Capital Stock

Housing is the best example of long-lived capital stock. Buildings last a long time, and it is often difficult to reinsulate them or to adapt their heating systems for different fuels. It is very costly, for example, to convert from electric resistance heating to more energy-efficient heating systems. Similarly, natural gas is available only where there are lines for it, so conversion from oil to gas is not always possible. The choices available to prospective purchasers and renters in a housing market are limited by the stock of existing housing; with high mortgage rates and construction costs, there is less building and the stock is replaced more slowly. A shortage of new buildings further limits consumers' ability to choose energy-efficient housing.

The Needs of Changing Households

Some demographic shifts affect both the patterns and the magnitude of energy use. With fewer persons in each household—the trend in the 1970s—per capita energy use generally rises even though smaller households may occupy smaller dwelling units. This is because certain energy expenditures are fixed regardless of household size (Abrahamse and Morrison, 1981). The increasing prevalence of dual-earner families has also changed energy demand. Wives and husbands may need to be in different places at the same time, prompting frequent trips by automobile (Abrahamse and Morrison, 1981). Also, because of the premium they place on their time, two-earner families may be more inclined than their one-earner counterparts to substitute energy for labor in housework through such labor-saving devices as dishwashers. The present trends have increased the need for housing units, automobiles, and appliances, as well as the demand for transportation during rush hours. These trends not only increase energy consumption, they also carry considerable inertia, since work patterns, living arrangements, and commuting patterns are highly resistant to change. Other demographic trends may decrease energy use, however, at least as compared with past rates of growth. Many major household appliances have nearly saturated the market, and the increasing number of elderly people in the population should bring decreased per capita energy demand, at least for transportation (Zimmerman, 1980).

Limited Access to Capital for Energy-Efficient Equipment

Usually, the most effective means of saving energy for residential consumers involve capital investments, while saving energy without using capital tends to involve the loss of amenities (Stern and Gardner, 1981). This is also likely to be true in the commercial and industrial sectors. As a result, users who wish to limit energy use must often choose between using capital and sacrificing comfort, mobility, or other values. Consumers who cannot afford to invest are forced to sacrifice.

The fact of limited access to capital has severe implications for low-income households because their economic need to improve energy efficiency is so great. Partly because these households tend to be in old and energy-inefficient housing, the proportion of household income they spend for energy is relatively high. One study found that people in the lowest 10 percent of the income distribution spent 30 percent of their income on energy for home and for transportation, while the median household spent only about 9 percent (Joint Economic Committee, 1977). More recent data show that the poorest households spend about 25 percent of their income on energy for home use, excluding transportation, while the richest households spend only about 2 percent (Energy Information Administration, 1982). Another study found that low-income households spend at least three-and-a-half times as great a proportion of their incomes on heating fuel as the average household (calculations from Community Services Administration, 1980), and the disparity is increasing. Between 1978 and 1980, home energy expenditures for the lowest-income households increased 48 percent; they increased only 17 percent for the highest-income households (Energy Information Administration, 1982). This economic pressure provides motivation while limiting the capital available for taking appropriate action.

The ability to make capital investments, either from saved or borrowed funds, is not only limited by income and assets, but is also affected by the policies of governments and lending institutions. Federal energy credits have lowered the eventual cost of energy-efficient equipment to home-owners and businesses, although not to the owners of rental housing. Tax credits, however, do nothing to make capital more readily available, nor do they provide a significant benefit to low-income households (Ferrey, 1981). Some utilities have sponsored low-interest and interest-free loan programs for energy efficiency, and these have had some limited success in getting capital to homeowners.

One likely implication of a scarcity of capital for energy efficiency is that different groups of energy users will respond differently to increasing energy prices. The more affluent are more able to make needed investments with little sacrifice of the services that energy provides. The less affluent have more limited options available, and so often respond with curtailments. Thus, a recent report (Oskamp, 1981) found that large chemical companies tended to invest in energy-efficient operations earlier than smaller companies, and another report the same year (Mills, 1981) found that large retail chains took more energy-saving actions than smaller chains.

Another example of segmentation is the fact that in the Northeast, homeowners have saved more energy by efficiency improvements than renters. (Stern, Black, and Elworth, 1982a). Both groups have curtailed energy use about equally by lowering indoor temperatures. But because homeowners have been able to save more overall, they have less often been compelled to fall behind in paying energy bills, to take extra jobs, or to make other economic sacrifices to pay energy costs. These experiences may be shifting the symbolic meaning of energy conservation in different directions for different groups and may also leave some groups with limited ability to respond to shortages of needed fuels.[8] Thus, the affluent may have greater ability to adapt in an emergency, in the sense that their past sacrifices have not been as severe.

Summary

Several important properties of the energy environment keep energy users from making the decisions their economic self-interest would dictate, even if they want to make self-interested decisions and even if accurate information is available. Energy users often do not know what responses are effective in their situations because energy and energy savings are invisible to them. Experience gives people good reason to distrust energy information, and they are likely to distrust the useful information as well as the misleading. People sometimes respond to the symbolic significance of energy, which is separate from its economic meaning. Finally, many of the important decisions about energy use have already been made by intermediaries, preempted by past decisions, dictated by demographic factors, or determined by limited access to necessary capital. Some of these properties of the energy system can be changed, and we offer some suggestions in the next two chapters. We also discuss energy users themselves, to provide further grounding for policy in an understanding of the human dimension of energy use.

Notes

1. Investments are considered economically justified when they provide the services energy is used for at the lowest possible cost to the users. This is only one of several possible criteria for an ideal

level of energy demand—one that derives from a view of energy as a commodity and a criterion of economic efficiency for the allocation of commodities. Criteria based on other views of energy would imply other ideal levels of energy use. For example, a concern with energy resources might lead to a desire to drastically curtail demand for *nonrenewable* energy sources, which account for over 90 percent of current energy demand, and a willingness to pay more for energy services to achieve that. Concern with national security might lead to a desire to decrease the demand for *imported* energy and a willingness to pay more for domestic energy—a national security premium. And concern with energy as a necessity could lead to a desire for an *increased* ideal level of demand if some consumers are believed to be deprived of energy services.

2. The relative overestimation of the importance of events that come frequently into awareness is an example of a well-documented general phenomenon in cognitive psychology (Tversky and Kahneman, 1974).

3. Of 443 written comments received on the very long and detailed regulations proposed in 1979 for the RCS, 159 discussed a single definition—that of "furnace efficiency modification." These comments reflected a conflict between electric and gas utilities. Many of the former argued that heat pumps should be considered as energy-efficient replacements for all types of furnaces; many of the latter argued that heat pumps should be considered as conservation improvements only when replacing electric resistance heating. The Department of Energy simply passed this controversy on to the states. Arguing that energy savings from fuel switching can vary greatly with climate, the department decided not to require energy auditors to consider fuel switching, but to allow states to institute such a requirement. (*Federal Register*, November 7, 1979, 44:64604).

4. The RCS had been a major target of pressure for regulatory relief, and its regulations were high on the list of those set for review by Vice-President Bush's Regulatory Relief Task Force in 1981 (see, e.g., Hershey, 1981; Berry, 1981). The result of the pressure for deregulation was a policy of eliminating all features of RCS that had not been specifically enacted into law by Congress (*Federal Register*, November 12, 1981).

5. The technical issues involved in comparing the energy use per passenger-mile of automobiles and passenger trains are rather complex, with the outcomes of the comparison dependent on a variety of assumptions about the technologies and the ways they

are used by passengers. However, during the Carter Administration, the conclusion that rail transportation is more energy efficient was accepted as fact. It was supported by a number of careful studies, and it was the conclusion of the Transportation Energy Conservation Data Book (Kulp, Shonka, Collins, Murphy, and Reed, 1980), published with the support of that administration. Nevertheless, when the Reagan Administration presented its first budget proposals for Amtrak, Transportation Secretary Drew Lewis referred to the energy efficiency of passenger trains as a "myth." (*New York Times*, March 11, 1981).

6. However, this reaction will not be universal. For low-income households and poorly supported public services, even a slow increase in energy prices will be experienced as a forced choice between sacrificing the services energy offers and sacrificing other essentials. The less painful options of cutting back excess or wasteful use and investing in energy efficiency are not available. This situation is already evident in the rapid rise in the last few years in the proportion of income that low-income households spend on energy (Energy Information Administration, 1982).

7. Unlike energy-efficient appliances, however, energy-efficient automobiles tend to cost less to produce because energy efficiency is gained primarily by decreasing the weight of the automobile, and, therefore, the cost of materials. Producers of such automobiles tend to gain a market advantage through lower price.

8. The notion of limited ability to respond to shortages rests on two assumptions: first, that curtailment is the most effective quick response in an emergency; and second, that there are limits to curtailment—such as the human physiological response to cold.

4

Individuals and Households as Energy Users

Individuals and households use about one-third of the nation's energy for space heating and cooling, transportation, and various household uses. In addition, they influence an even larger portion of energy use indirectly through their purchase decisions, which partly determine the amount of energy used in producing consumer goods and services. To understand energy demand, one must understand the energy-using behavior of individuals and households.

This book approaches understanding energy users differently from most previous studies. As we have noted, most past analyses have derived from a conception of energy as a commodity. Since energy use, in this view, is a type of consumer behavior, it is no surprise that most analyses of energy use have applied the dominant theory of consumer behavior—the theory of rational choice. These analyses have assumed that energy users—both individuals and organizations—are "rational"; that is, that they act in their own self-interest to maximize some objective function. The most common assumption about individuals and households is that they act to maximize the value of consumer goods and services acquired within their budgets. The most common assumption about firms is that they, or their managers, act to maximize profits. In the public sector, the most common assumption is that agencies try to maximize size, which means number of employees, number of programs or offices, and budget. For a "rational" decision maker, energy decisions are like any other decisions. Energy conservation will occur when a person, firm, or agency expects to save more than a dollar per dollar spent.

Even in this dominant view of energy consumption, it is acknowledged that energy users do not always take the actions that will benefit them most. Such behavior is usually attributed to short-term energy price fluctuations, to time lags in adjustment, to the unavailability of complete information to guide decisions, or to impediments to market functioning, such as the presence of price controls on energy, the existence of regulated utilities, and the prevalence of situations such as rental housing, where purchasers of efficient energy-using equipment do not benefit from the investment.[1] It follows from this argument that government action can make the energy system more efficient by removing impediments to market functioning, providing information, or offering incentives or penalties through tax policy, assistance programs, or through regulation. Such policies can shorten the time it takes for energy users to take actions that most benefit them.

Despite some evident differences in energy policy among recent federal administrations, all have operated on the underlying assumption that when individuals or organizations use energy, they are making rational economic decisions aimed at maximizing some objective function. In the Ford and Carter administrations, this view provided the rationale for programs to inform citizens of the energy costs of major purchase decisions. It lay behind the Carter administration's tax incentives to speed conversion to energy-efficient operation of homes and businesses and it helped justify the removal of oil price controls by the Carter and Reagan administrations. Such diverse governmental actions as low-income weatherization assistance, small "appropriate technology" grants, energy performance standards for buildings, and even the elimination of these same programs have all been justified in terms of the assumption that energy users make economically rational decisions.

While it may seem strange that the same basic assumption has been used to support opposing policies, the assumption remains useful for predicting and interpreting aggregate changes in energy use, and it has practical implications for policy. For example, the simple assumption of rationality correctly predicts that when oil prices rise relative to the prices of other fuels and of energy-efficient equipment, some energy users will switch from oil to other fuels, and some will invest in energy-efficient equipment. To cut oil use, then, the assumption of rationality suggests raising oil prices, and data from the United States and other industrialized nations show that this policy is effective (e.g., Marlay, 1982; Schipper and Ketoff, 1982).

Careful observation of individuals and organizations provides support for the importance of simple cost and return factors in the behavior of energy users. However, careful observation also makes it clear that other factors are involved. For example, in Chapter 3, we described how energy users may lack the knowledge to take advantage of information conveyed in energy prices and how they fail to act on information they distrust.

There is other evidence that a variety of social, political, economic, and personal influences are significant determinants of energy consumption.[2] It has become a commonplace observation, for example, that different families of the same size occupying identical residences can vary in their energy use by a factor of two or more (Lundstrom, 1980; Sonderegger, 1978). Thus, the behavior of building occupants is often a major factor in the building's energy consumption. Even more compelling is the fact, demonstrated in numerous careful studies, that energy-using behavior can be altered greatly while technological and economic factors remain constant. We describe here some examples related to residential energy use.

Michael Pallak and his colleagues (Pallak, Cook, and Sullivan, 1980) showed that, all other things being equal, getting people to pay attention to their energy use led them to reduce consumption. They asked seventeen Iowa homeowners to participate in a study aimed at determining whether personal efforts could make much difference in saving energy. The researcher asked the homeowners to keep an "energy log" by noting their appliance use twice a day and reading their electric meters weekly. At the end of a month, these homeowners were using 13 percent less electricity than a group of sixteen control households that had agreed to participate but had not been asked to keep an energy log. The experiment officially ended at that point, but among the experimental households, the energy savings continued for almost a full year.

More aggressive efforts to influence energy use make a bigger difference. Richard Winett and his colleagues (Winett, Hatcher, Fort, Leckliter, Love, Riley, and Fishback, 1982) combined knowledge of behavioral psychology and communication techniques into a carefully constructed package to teach and motivate householders to cut energy use in their all-electric apartments and townhouses without spending money on equipment and with minimal loss in comfort. Their experimental program featured twenty-minute videotape programs that showed a young couple acting as a model by taking energy-saving actions in their home. The videotape on summer energy savings, for example, demonstrated the proper use of fans and natural ventilation in the evening; ways to shift the time or place of activities such as cooking and eating to decrease the need for air conditioning; dressing in lightweight clothing; and so forth. The script was carefully constructed to present energy efficiency as a positive action rather than emphasizing conservation. Participants in the study all attended a forty-five-minute meeting explaining the project at which they were given instructions in the proper use of window fans, information on the exact insulating value of different items of clothing, and information on how to use a hygrothermograph installed in their homes to monitor temperature and humidity. In addition, some of the participants were given daily feedback for thirty days on the amount of energy they were using.

The group that saw the videotape cut total electricity consumption by 10 percent in comparison with a control group that only attended the meeting. In a three-week follow-up after the experiment, the savings were 19 percent. These savings amounted to 26 and 63 percent, respectively, of the electricity used for air conditioning. This saving was accomplished with little or no change in indoor temperature and no change in residents' comfort. When householders were given daily feedback on their rates of energy use, in addition to the videotape, their savings increased even further. In a parallel experiment with winter energy savings, the combination of videotapes and feedback produced a total savings of energy of more than 25 percent of the electricity used for heating.

These studies demonstrate that social and psychological factors can make a sizable difference in residential energy use even when both economic incentives and the physical properties of the building are held constant. Of course, energy use depends not only on the habits of building occupants, but on levels of investment in insulation, energy-efficient heating and cooling equipment, and other practical devices. The evidence is that most of the remaining potential for energy savings in the residential sector requires such investments.[3]

Furthermore, the level of investment in energy efficiency can be increased substantially by changing social and psychological conditions—with very little reliance on special financial incentives or penalties. For example, Stern, Black, and Elworth (1981, 1982a) studied a program in the Northeast that offered homeowners a combined package of home energy audits, assistance with financing, contact with certified contractors, and inspection of energy conservation work done on the home. The program was financed in part by a surcharge on the work done under contract, so the program did not provide its services at the lowest available cost to consumers. Nevertheless, 2,000 households—about one-quarter of those who requested its free energy audits—went on to have the recommended work done by the program. These homeowners made investments that will save them almost four times as much energy as will be saved by the investments of homeowners who did not participate in the program. In fact, even participants who received free energy audits and declined to have work done through the program reported making investments that will save twice as much energy as the comparison group.[4] The most frequently given reasons for signing contracts with the program were distinctly nonfinancial: "I trusted the work because it would be inspected" (98 percent); "I didn't have to worry about finding a reliable contractor" (96 percent); "The staff was very professional and trustworthy" (89 percent); "It was convenient to have them do the recommended work" (62 percent). In short, the program's qualitative advantages, rather than simply the net financial benefit to be expected from energy efficiency, made a large difference in household behavior and in energy savings.

Each of these studies demonstrates, in a different way, that nontechnical and noneconomic factors can have a major impact on energy use. Taken together, the studies suggest that the human dimension may explain a significant proportion of what has been unknown about energy use in the United States. The findings showed that householders' behavior is not readily predicted from the notion of rational decision making. In Pallak's research, energy was saved when people were induced to pay attention to information that was already available to them with little effort. Winett's procedures combined new information with motivational techniques, fairly sophisticated use of media, and again, attention to information that was already available. The study by Stern and his colleagues showed that certain nonfinancial but significant features of a conservation program might make more difference to householders than an interest subsidy or other financial inducements. In none of these cases was household behavior "irrational," but these studies showed that people often do not act in their economic self-interest, despite the availability of information sufficient for such action. The studies further suggest that there is considerable practical potential for residential energy savings without modifying existing economic incentives. Much of this potential can be realized by building on an empirically based knowledge of what actually is keeping energy users from taking actions that will benefit them.

In the next section, we define five different views of energy users, each of which is supported by behavioral knowledge. This knowledge is then applied to our previous analysis of barriers to energy efficiency (Chapter 3) and to the problem of providing energy information to individuals and households. We identify some principles and offer some concrete suggestions for making energy information effective, and we present a detailed discussion of home energy audits as an example.

FIVE VIEWS OF THE INDIVIDUAL AS ENERGY USER

Energy User as Investor

Energy users can be regarded as investors for whom energy has a cost that is carefully considered in making purchases of equipment that uses energy. User-investors consider such equipment as capital, in the sense that it is a durable good that produces a stream of economic benefits, such as reduced energy costs, over its useful life. Building or vehicle owners may also see the purchase of energy-saving equipment as an investment if they expect it will increase the resale value of a property.

The view of an energy user as an investor is completely consistent with the assumption of economic rationality. Every individual may be seen as acting to maximize future disposable income. In theory, investments are based on stable preferences and on an analysis of the discounted future value of energy expected to be used or saved. A family decision to exchange a large "gas guzzler" automobile for a new fuel-efficient subcompact car can be regarded as an investment decision: for a capital investment of, say, $7,500 less trade-in, the family will be able to reduce its energy costs by a predictable annual amount. This sort of analysis underlies the practice of analyzing expenditures on energy-efficient technologies in terms of payback period or return on investment.

There are many ways an energy user might calculate the expected outcome from an energy investment. Economists generally argue that the best, most accurate index is the internal rate of return. This index is the interest rate that would make the present value of the stream of benefits expected from the investment equal to the initial cost of the investment. It is considered best because it takes into account the fact that a dollar now would grow if invested, and because rate of return allows easy comparisons with the value of alternative investments. An internal rate of return, however, is difficult to calculate, since it is the sum of a mathematical series. It requires careful mathematics and uncommon patience—or a small computer. In fact, many economic analyses of energy-efficiency investments have used more simplified indices, such as the present-value, time-discounted cost-benefit ratio or the payback period. If the most accurate index of the value of an investment is difficult even for economists to use, it is not surprising that few ordinary energy users do these calculations.

Individuals tend to quantify most household energy sources in dollars, rather than in energy units (Kempton and Montgomery, 1982). This difference in estimation procedures makes energy users behave very differently than an expert's analysis would predict. Figure 3 shows two sets of calculations of a simple index of investment—the payback period from an investment that costs the equivalent of one year's fuel cost and that cuts energy use by 30 percent. A payback period is the time it takes to recover the cost of investment through energy savings. The "expert model" shows that the initial cost of the investment is paid back faster if fuel prices increase, because more costly fuel is being saved. The "folk model," by contrast, calculates savings in dollars compared to preinvestment expenditures. In this model, fuel price increases can quickly make a 30 percent fuel savings disappear because fuel bills return to their preinvestment levels. While this folk model may be demonstrably "irrational" in economic terms, it does follow logically from the method most commonly used by individuals to judge the effects of attempts to save energy. People who try to make rational calculations based on their own assumptions about energy would be led to make fewer energy-saving investments than an expert analyst would recommend. Not only would they interpret their investments as less effective than would an expert; they would also communicate this judgment to their friends. It is of little use to decry the unsophisticated

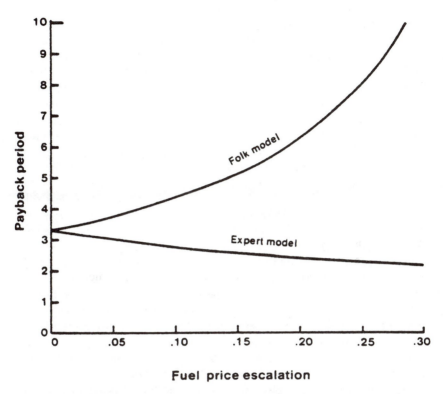

Fuel price escalation

Fig. 3. Payback period as a function of fuel price escalation, as computed by a folk model and an expert model
SOURCE: Kempton and Montgomery (1982)

calculations made by the folk model: it approximates the ways most individuals calculate, if they calculate at all.

More fundamentally, there is a problem with the very notion of users as investors. People do not see their purchases of energy and energy-using equipment only as investments; they have meanings unrelated to the cost of fuel. Car purchasers, for example, do not look solely at fuel efficiency. They are also concerned with performance, safety, styling, status considerations, and other factors. To take another example, decisions about home improvements can have major implications for household energy use. The homeowner may view these decisions as economic investments, in the sense that home improvements may have a continuing benefit by reducing operating costs, but they also have implications that do not easily translate into return on investment. They may increase comfort, provide more space or light, or improve the appearance of the home. Thus, when a homeowner considers reinsulating or replacing a working furnace, that choice is competing against unlike alternatives—another bathroom, a picture window, new living-room furniture, and so forth. People do not usually weigh the potential value of the energy saved by one purchase against the pleasure, convenience, or status achievable by alternative purchases. People are not likely to treat energy efficiency strictly as investment when they are not likely to consider the alternatives to energy efficiency as investments.

In households, other important energy decisions are made when furnaces, water heaters, refrigerators, and other equipment wear out. Such decisions may be made under time pressure and with only partial information, with quick replacement a more pressing issue than life-cycle energy costs. Although those decisions may affect energy use for years or even decades, the purchaser may see them as repair, not investment, decisions. Seen as a repair, a $200 water heater will be viewed as costing less than a $300 water heater—even though the latter is better insulated and consumes $25 less in energy each year. In addition, there are the serious difficulties that are involved in any effort to find complete information.

As already noted, many important decisions about energy use are made by intermediaries who are not the ultimate users of energy-using equipment. These intermediaries include developers of residential and commercial buildings and operators of automobile rental agencies. In these cases, the investors do not pay for the energy used. The investors' concerns have to do with ultimate sale or lease of a product that is competing with similar products. For example, in new multihousehold residences, electric resistance heating is often installed to keep down the price of construction of the building, to shift responsibility for heating to the occupants, and to let the building owner escape the various management problems associated with central heating. Decisions about what appliances to install in a new building are often based more on visual appeal than on life-cycle cost. While these are certainly investment decisions, future energy costs are not involved since they will be paid by someone other than the person who makes the investment decision. Thus, the investor is relatively unconcerned with energy consumption, and the energy user is uninvolved in purchasing the durable goods that might be seen as an investment.

For all these reasons, to view energy users as only investors leads to an inaccurate account of individual behavior, even with respect to capital goods. Other views of energy users are often more applicable.

Energy User as Consumer

In another view, individuals think of their homes and automobiles as consumer goods, that is, as providing necessities and pleasures. Energy-using activities and equipment are purchased for the value of using them. Once purchased, they require money primarily to maintain or increase their ability to provide necessities and pleasures and only secondarily to increase their economic value. This view might offer an explanation of the fact that home improvement loans are taken out much more frequently for room additions or new siding than for reinsulation or new and improved furnaces. The former expenditures give owners pleasure or tangibly add to a home's appearance, in addition to their economic benefits; the latter are unseen and mainly save money. Among energy investments, the usual preference for storm windows rather than wall insulation may reflect the same phenomenon: storm windows are attractive, cut down on street noise, and may decrease the physical effort of home upkeep; insulation offers a faster return on investment, but it lacks these consumer benefits.

This view of the energy user can be consistent with an assumption of rational action if people are assumed to act to maximize some subjective quality—what economists call a utility function. The usefulness of an assumption of rationality for predicting behavior would then depend on empirical knowledge of such utility functions and on a demonstration that they are based on reasonably stable preferences. Such evidence is lacking, so the view of the energy user as consumer is described here as a heuristic rather than a formal model.

Energy Consumer as Member of a Social Group

Homes and automobiles also may have social meaning. They express membership in a community or attainment of a certain status in society. In youth, the keys to the family car symbolize attaining adult status; a home in the suburbs often symbolizes career success. And that home must be acceptable in appearance to the occupant's friends, neighbors, or co-workers—or the homeowner risks loss of status and rejection. As a local energy manager remarked to a member of the committee: "I'm committed to saving energy and I know plastic sheeting over my windows would have a fast payback, but I wouldn't dream of putting plastic on my house. My neighbors would kill me." Energy considerations almost always take second place when they are in conflict with strong social pressures.

Social group memberships are also important as sources of innovation and of energy information. A homeowner may get the idea to install a clock thermostat from seeing one in a friend's or neighbor's home; the decisive information about whether the investment is a good one may come from the experiences of that friend or neighbor. When this happens, the action is most accurately described by the metaphor of social contagion, even if the individual rationalizes his or her action in terms of expected financial return. Since such action is not the outcome of detailed search for information and may not produce the maximum expected benefit, it is not rational in the formal sense. It may not even approximate formal rationality—individuals may rely on sources that can add no accurate information whatever. Still, reliance on word-of-mouth information from friends or associates may be a sensible strategy under some circumstances, such as when more formally prepared information is conflicting and untrustworthy.

Energy Consumption as Expression of Personal Values

Individuals use or conserve energy in ways consistent with their personal ideals or their self-images. For some, central air conditioning may be an important expression of a value of comfortable and gracious living. For others, solar collectors on the roof may express values of self-reliance or environmental preservation.

There is evidence of important variations in energy-related values. When presented with a choice between energy and environmental values, for example, women and younger people usually express greater preference for environmental protection than men and older people (Farhar, Weis, Unseld, and Burns, 1979). This may be significant because environmental concern has an effect on energy-related behavior (Black, 1978; Stern, Black, and Elworth, 1982b, 1983; Verhallen and van Raaij, 1981).[5] The age difference with respect to environmental values may be particularly important in the future. If environmental concern among the young is a reflection of the increased importance of environmental issues during their formative years, the data may portend increasing importance of environmental values as an influence on energy consumption.

Energy User as Problem Avoider

According to the view of energy users as problem avoiders, people usually take energy use for granted and treat it as no more than a potential source of annoyance or inconvenience. Nothing is done about energy until the furnace breaks down, a power outage or a gasoline shortage occurs, or there is such a sharp rise in the price of energy as to command immediate attention either because of the change itself or because energy becomes a more significant portion of the budget. In this view, attention is a scarce resource. People do not change their energy-use patterns until some threshold of annoyance is passed. At that point, they respond, and that behavioral response continues until some new and pressing problem appears to change behavior again. Behavior typically is haphazard and oriented toward short-term avoidance of inconvenience, perhaps guided by hearsay, rule of thumb, "what worked last time," or other unsystematic influences.

This view implies that people will not take energy-saving action if this action involves significant inconvenience or disruption to household routines. A paper by Penz (1981) contains a detailed anecdotal account of the process of deciding about home insulation and is convincing on this point. Penz followed numerous false leads from the telephone *Yellow Pages* and other sources, spoke with unresponsive retailers and utilities, and went through the exhausting process of calling unknown contractors and comparing their recommendations, their bids, and his impressions of their competence. The prospect of all this effort, no doubt, is sufficient deterrent for many homeowners.[6]

Householders sometimes try to avoid problems like this by a strategy of relying on a single trusted source for energy advice and services. But reliance on a heating oil delivery company, a furnace repairman, or a handy neighbor may not be the best way to get energy services at the lowest cost because the trusted source may not be expert in the relevant area. Worse, the source may have an interest that conflicts with the energy user's. But the alternative of searching for accurate information and reliable service providers in the present energy environment is perceived as an extremely onerous task. As a result, many people are willing to trade the likelihood of saving on some energy costs for the sense that they have done something to improve their situation and the assurance that they are avoiding a major loss or ridding themselves of an annoyance.[7]

A good example of this consumer strategy can be found in the study by Stern and his colleagues (1981), already mentioned, of a comprehensive residential conservation program in the Northeast. Among program participants, the reasons most often given for signing contracts with the program reflected issues of risk avoidance and convenience. These reasons are most readily interpreted in terms of a motive to avoid problems, especially the very costly ones that may result from unsatisfactory work by contractors.

The notion of avoiding problems also includes difficulties that arise among household members. It is tempting to think of a household acting as a unit, but this view is not always accurate. If an effort to cut energy use involves arguments about thermostat settings, or negotiation over who will use an automobile, or nagging children to turn off appliances, take short showers, and the like, a great many people will quickly revert to their previous patterns of energy use.

SOCIAL AND PSYCHOLOGICAL PROCESSES AFFECTING ENERGY USERS

The five views of energy users are presented not to argue that one is superior, but because there are elements of truth in each. It is appropriate to consider the conditions under which each view furnishes useful insights about how energy users behave. The following discussion elaborates on some of the processes that lead energy users to behave as other than investors.

Interpersonal Influence and Imitation

We have already described the environment in which energy users function. In many ways, it is the sort of environment in which interpersonal influences are likely to have especially strong effects on individual behavior. In particular, energy prices have been changing rapidly and unpredictably, creating a high degree of uncertainty. Decades of experimental research have demonstrated that in such an uncertain environment, the ideas and actions of other people have a heightened effect on individual judgment and behavior (e.g., Latané and Darley, 1970; Sherif, 1935). We have also noted that information about energy is conflicting, and many of the usual sources of information are not generally trusted. In such an information system, expertise is seen as unreliable, and the need for credible information leads people to reliance on trustworthy sources, even if they are less expert.

Given this confusing environment, it is to be expected that social influences, especially the influence of peers, will be important determinants of energy use. The view of energy users as members of social groups should have increased applicability here, and data about influences on behavior in social groups will be relevant to energy use.

Imitation.   Simple imitation is one form of influence. An illustration is provided by Aronson and O'Leary (1983), who demonstrated in a university building how this form of influence can induce people to modify a routine habit—taking a shower—in a manner that conserves energy. Even though an overwhelming majority of students using the shower room knew that taking short showers saves energy—and even though a prominent sign on the wall reminded them to keep showers short and to turn off the water while soaping up—only 6 percent of the students took the recommended steps to conserve hot water. When the researchers made the sign obtrusive, short showers increased to 19 percent. But many students were offended by the intrusiveness of the sign: some expressed anger verbally, a few knocked over the sign, and some expressed their hostility by taking unusually long showers. The researchers next used students who served as models by turning off the water and soaping up whenever someone came in to use the facility. When this strategy was used, the number of people turning off the water to soap up climbed to 49 percent; with two people simultaneously modeling the behavior it rose to 67 percent.

Imitation can be a potent influence even on more significant energy-related activities. As we already noted, videotaped demonstrations by people acting as models of energy-saving behavior were a central feature in the highly effective experimental program developed by Winett and his associates at Virginia Polytechnic Institute (Winett et al., 1982; Ester and Winett, 1982). Their data indicate that information alone is of limited value; imitation was crucial to the magnitude of the effect.

Simple imitation has its limits, of course, even when the influence of models is made more available through videotape technology. Although imitation has proven effective with simple behaviors, it seems unlikely that a videotape program will lead many to adopt such expensive measures as installing insulation in walls or purchasing fuel-efficient water heaters. These sorts of actions are more likely to be influenced by direct communication with friends and associates.

Communication. The power of influence by interpersonal communication, rather than by experts or media appeals, has been documented for a great many years. Consider the history of the agricultural extension program in the United States. During the 1930s the federal government attempted to disseminate information about improved agricultural practices. At first the government tried to persuade farmers by distributing pamphlets filled with important information in the form of tables, charts, and statistics. This information campaign was a dismal failure. Subsequently, a demonstration project was set up in which government agents worked side by side with farmers on a few selected farms. When neighboring farmers saw the size of the demonstration harvest and discussed the methods that were used, they quickly adopted the new techniques (Nisbett et al., 1976).

This sort of personal influence process has been documented repeatedly and has recently been shown to operate among energy consumers. In studies of residential adoption of energy-conserving practices and of solar energy equipment, Dorothy Leonard-Barton (1980, 1981a) found that interpersonal sources of communication are considered most important both by adopters and nonadopters of the equipment. This finding is typical: while media sources may be most effective for letting people know a new technology exists, personal sources are more influential in the decision to adopt the innovation (e.g., Katz, 1961). In Leonard-Barton's research, the best predictor of intention to purchase solar equipment was found to be the number of solar owners that a potential adopter knows. This finding exemplifies the general principle that certain people—particularly those with knowledge and experience and extensive social contacts—act as informal opinion leaders to influence others (Katz and Lazarsfeld, 1955). In a study of the adoption of energy conservation equipment, Darley (1978) found that the adoption of a newly developed clock thermostat spread from the people who first used it to their friends, colleagues, and co-workers. The spread was along lines of communication, not through mere physical proximity; neighbors were not usually the next to try the new equipment.

The experiences of family, friends, and colleagues are influential for several reasons. First, as we have mentioned, the uncertainty in the energy environment and the lack of trust in formal information sources enhance the influence of people who are well known and trusted. Friends are trustworthy sources of information; at the least, their biases and values are known, and can be taken into account.

Second, when a friend or colleague adopts some energy innovation, for example, a solar water heater or the practice of car pooling, that adoption represents a vicarious experiment for the person who sees or hears about the effort and its results. Research on the diffusion of innovation demonstrates that people are more likely to adopt a new idea or technology if they can try it on a small scale without fully committing themselves to it (Rogers with Shoemaker, 1971); a friend's experience can act as such a trial.

Third, information from close associates is salient—it stands out from the mass of available information and attracts attention. Part of this effect of salience (Taylor and Fiske, 1978) is due to people weighing information in proportion to its vividness (Nisbett et al., 1976; Borgida and Nisbett, 1977; Hamill et al., 1980). The experience of a friend or acquaintance may yield a vivid demonstration, not just a vicarious experiment, of what one might expect from adopting the innovation oneself. Impersonal data summaries, even from large numbers of cases, have been shown to have less impact than vivid face-to-face interactions and detailed case studies. Vividly presented information stands out from other information and is more likely to be noticed, remembered, and given weight in judgments (Taylor and Thompson, 1982).[8] The experience of someone one knows well, combined with the opportunity to hear the experience firsthand, exerts an influence far greater than its status as additional information. This is true even if the friend's experience is atypical.

A study by Nisbett and his colleagues offers this example (Nisbett, Borgida, Crandall, and Reed, 1976:129):

> Let us suppose that you wish to buy a new car and have decided that on grounds of economy and longevity you want to purchase one of those solid, stalwart, middle-class Swedish cars—either a Volvo or a Saab. As a prudent and sensible buyer, you go to *Consumer Reports*, which informs you that the consensus of their experts is that the Volvo is mechanically superior, and the consensus of the readership is that the Volvo has the better repair record. Armed with this information, you decide to go and strike a bargain with the Volvo dealer before the week is out. In the interim, however, you go to a cocktail party where you announce this intention to an acquaintance. He reacts with disbelief and alarm: "A Volvo! You've got to be kidding. My brother-in-law had a Volvo. First, that fancy fuel injection computer thing went out. 250 bucks. Next he started having trouble with the rear end. Had to replace it. Then the transmission and the clutch. Finally sold it in three years for junk.

If the data in *Consumer Reports* were based on 1,000 cases, the information you received at the cocktail party has now increased the sample to 1,001 cases. But people do not respond to this event according to its logical statistical status. Rather, the single event often has a decisive impact far beyond its logical status.

Changes in energy use are affected by the role of friends and acquaintances in spreading new ideas in an uncertain environment. New energy ideas may be held back at first because the fact that they have not been tried is taken as evidence that they do not work. But on the positive side, once a new method of saving energy has been tried, it can spread easily along predictable channels. Word of mouth is a particularly important medium of communication for social groups that either do not trust information from established institutions, or do not receive the information transmitted in other media due to lack of access or to language problems.

The Momentum of Past Behavior

Behavior once undertaken often requires additional bolstering and justification that in turn leads to a shift in values. People who have recently made an important decision seek to justify that decision after the fact— convincing themselves and others that the decision was a wise one. This behavior is predicted, explained, and researched under the rubric of the theory of cognitive dissonance (Festinger, 1957; Festinger and Aronson, 1960; Aronson, 1969, 1980; Wicklund and Brehm, 1976).[9]

A few general findings and principles with potential relevance to energy use have come from this research. One, people tend to rationalize the choice they have made in a difficult decision. They tend to emphasize the positive aspects of the chosen alternative and the negative aspects of the unchosen alternative. As a result, as time goes by, the individual comes increasingly to view the selected option as clearly superior to the unselected one (Brehm, 1956; Darley and Berscheid, 1967). Two, the greater the commitment in terms of effort, cost, or irrevocability, the stronger and more permanent the effect (Aronson and Mills, 1959; Axsom and Cooper, 1980; Gerard and Mathewson, 1966; Knox and Inkster, 1968). Three, people tend to remember the plausible arguments favoring their own position and the implausible arguments opposing their position (Jones and Kohler, 1958); this serves the need for self-justification rather than that of objective fact-seeking. Four, once someone makes a small commitment in a given direction, that person is much more likely to make a large commitment than someone who is uninvolved (e.g., Freedman and Fraser, 1966).

Applied to energy consumption, these principles describe an inertia in behavior: people resist change because they are committed to what they have been doing, and they justify that inertia by downgrading information that implies that change is essential. This partly explains the failure of many energy users to take economically justifiable action to save energy. But these principles also suggest that change may be brought about by a process that begins with small commitments to energy-saving action and then moves under its own momentum toward more significant efforts.

These principles have been demonstrated in controlled field experiments. Pallak, Cook, and Sullivan (1980) applied principles of commitment and self-justification directly to energy conservation. They reasoned that even when people are convinced that a particular course of action is desirable for themselves or the community, they still need a little help in overcoming inertia. "I really want to . . . [donate blood, get flu shots, give up smoking, go on a diet, conserve energy,] but . . . [I can't find the time, I'll start next week, etc.]" The researchers started with a group of homeowners who volunteered to try to save energy by turning down their thermostats, wearing sweaters, taking shorter showers, and so forth. The experimenters randomly assigned the volunteers to one of two groups. Both groups were given the same information about energy conservation strategies, but one group was informed that the researchers hoped to list participants' names in an article about the experiment—thereby creating a high-commitment situation—while the other group was explicitly assured of anonymity. There was an immediate effect: people in the high-commitment group used about 15 percent less natural gas than people in the low-commitment group. In an identical experiment with electricity use, the difference between the two groups was close to 20 percent. More importantly, there was a lasting effect. After one month of observations, the volunteers were informed that the project had been successful in saving energy. But the homeowners in the high-commitment condition were told it would not be possible to use their names. The researchers continued to read the meters for the next eleven months and found that the homeowners in the high-commitment condition continued to use less energy than those in the control groups, although there was some decline in the magnitude of the difference as time passed. These findings have been confirmed in a similar, more recent study (Katzev and Johnson, 1982). The results of these studies are extremely provocative: they suggest that once a person believes he or she is publicly committed to saving energy, he or she adopts behaviors that can last much longer than the public commitment itself.

A recent study by Bruce Hutton (1982) evaluated the effectiveness of three advertising campaigns aimed at motivating energy users to adopt conservation measures and purchase energy-efficient products. Two of the campaigns emphasized television, radio, and magazine advertising in attempts to get householders to give more consideration to energy efficiency in their appliance purchases. These campaigns increased awareness, but they were ineffective at changing behavior. Furthermore, the response to the campaigns did not follow the usual expectation that "more is better": that is, more frequent exposure to a message did not mean greater impact. The only one of the three campaigns that produced a significant impact on behavior did not involve repeated exposure. Instead, an informational booklet and a flow-restricting shower head insert were delivered to each of 4.5 million households. The households often inserted the flow restrictor and went on to take other advice offered in the booklet—making furnace adjustments, insulating duct work, and so forth. These changes in behavior seem to have been due not only to the information in the booklet, but also to the momentum caused by installing the flow restrictor. Once a person makes a small commitment in the direction of energy conservation, his or her tendency to try other behaviors—especially if they are clearly described, inexpensive, and relatively easy—is increased.

These experiments present clear, energy-relevant examples of a general behavioral dynamic: (1) people profess a desire to make a change; (2) the degree of change they make is enhanced by an intervention that increases the degree of cognitive commitment to the change; and (3) it may be inferred that people are pleased with the outcome, or at least they did not experience coercion. In the experiments by Pallak and his colleagues and by other researchers, people continued to show significant behavior change long after the precipitating event had passed and the presence of the researchers had lost its salience. Other studies suggest that adoption of one energy-saving practice led easily to the adoption of others. Some of the practical implications of this dynamic are explored further in the section below on energy information programs.

The Expression of Personal Values and Norms

Energy use is influenced by broad personal values and by specific norms for action. Those values and norms are products of upbringing, perceptions of world and local events, and the influence of other people. They can take on the psychological force of moral convictions or of ego involvement. For example, a person who grew up in poverty may define his or her personal worth in terms of consumerism. Such a person may feel he or she has "made it," or succeeded in climbing out of impoverished circumstances, because there is no need to be concerned with cost or to worry about waste. "Why should I turn my thermostat down at night? I can afford it." That kind of attitude, based as it is on a person's sense of self, is a formidable barrier to energy-conserving actions. Similarly, a corporate executive is likely to feel that a person in his or her position should travel by air, ride in a large private car, and work in a spacious, climate-controlled office.

A very different value, which seems to be gaining favor among U.S. consumers, has been called "voluntary simplicity" (Elgin and Mitchell, 1977). Leonard-Barton (1981b) describes it in terms of a syndrome of behaviors: using bicycles for transportation, recycling paper, cans, and glass, learning to increase self-sufficiency, eating meatless meals, buying secondhand goods, and making certain items at home instead of purchasing them. In three studies in California, it was shown that people who score high on an index of these behaviors were more apt than others to intend to purchase, or to have purchased, wall insulation, furnace timers, and solar water heating equipment. They were also more likely to report such energy-saving behaviors such as turning off furnace pilot lights during summer months and weatherstripping doors and windows.

At least three types of individuals have been found to adopt lifestyles of voluntary simplicity. Leonard-Barton called them "conservers," "crusaders," and "conformists." The conservers are people for whom conservation is a long-established pattern of living that emphasizes thrift. Crusaders are motivated to live a life of voluntary simplicity by a strong sense of social responsibility—a conviction that everyone lives in a small, delicately balanced ecosystem whose resources need to be carefully husbanded. Conformists are people who engage in voluntary simplicity for less well-defined reasons: some seem to be motivated by guilt for possessing an excess of wordly goods; others are influenced by living in a highly energy-conserving and ecologically aware neighborhood. Such people are likely to revert to a high consumption lifestyle if they return to a social setting in which consumption is encouraged (Leonard-Barton and Rogers, 1979). If voluntary simplicity gains acceptance (Olsen, 1981), there are broad-ranging implications for energy use.

Specific energy-saving actions can also become associated with widely held values, such as helping others. When this happens, people feel a moral obligation to act. This process has been studied in a four-year field experiment with time-of-day pricing of residential electricity in Wisconsin. Utility customers were experimentally assigned to a variety of electricity rate structures that included a large jump in rates during the peak demand periods on weekdays. Participants who believed that lowered demand in peak periods would be good for people in general—for example, by allowing utilities to shut down inefficient and polluting power plants—and also believed that households as a group could make a big difference in peak demand, felt a moral obligation to lower electricity use in peak periods (Black, 1978). This feeling of moral obligation proved to be more important than price: people who felt an obligation to change their behavior had lower electric bills than people who felt no moral obligation, but who faced the same electricity rates. In fact, this effect was greater than that of price even when the price differentials between peak and off-peak hours ranged as high as 8 to 1 (Heberlein and Warriner, 1982).

General attitudes do not always predict energy use (Farhar, Weis, Unseld, and Burns, 1979; Olsen, 1981). But it is reasonable to suppose that personal values and norms will affect behavior unless action is constrained by other influences. A value of voluntary simplicity or a sense of moral obligation to use energy resources efficiently may provide an impetus for action, but some actions are easier to accomplish than others because of limitations on behavior imposed by the environment. People may be prevented from acting on their values and norms because decisions have been made for them by intermediaries; because they do not have the right to act, for example, by insulating the walls of a rented apartment; because they cannot afford some actions; or for other reasons. Thus, personal values and norms are more influential with respect to some energy-related behaviors than others.

There is evidence supporting this analysis. For example, Stern, Black, and Elworth (1982b, 1983) have classified residential conservation actions as either major capital investments, low-cost efficiency improvements, minor curtailments, or temperature setbacks. They attempted to account for each type of behavior in a statewide sample of homeowners and renters in Massachusetts. They found that personal norms supporting energy conservation were reliable predictors of temperature setbacks, a cost-free action available to almost all households. Norms were more weakly related to low-cost actions, such as weatherstripping, and to minor curtailments, and they showed the weakest relationship of all to major household investments, such as insulation and replacement of inefficient furnace equipment. In short, a personal norm supporting energy conservation is most likely to be converted into action if the action involves little cost in time or money.

In theory, this norm-activation process could be used by educational programs. Numerous controlled experiments with exhortations to save energy, however, have had generally unimpressive results (Ester and Winett, 1982; Stern and Gardner, 1981).[10] These suggest that it is not easy to shape norms with exhortation. Over time, however, the effect of personal norms—especially if they are changing—is almost certain to accumulate. When that occurs, an understanding of the relationship of energy use and conservation to individual values and norms will prove useful for forecasting trends in energy consumption. Understanding energy-related values and norms will also be useful for understanding public support and opposition to energy policies and programs.

MAKING ENERGY INFORMATION PROGRAMS EFFECTIVE

Chapter 3 and the previous material provide a basis for understanding household energy users. We now use the analysis to examine the problem of providing energy information to individuals and households. The home energy audit is discussed in detail at the end of the chapter as an important type of energy information program.

Federal, state, and local governments have devoted considerable effort to providing more complete and accurate information for individual energy users. Government agencies have demanded, developed, and disseminated a wealth and variety of energy information. Fuel economy tests have been mandated for automobiles, and the results are publicized in *Fuel Economy Mileage Guides*, which are distributed through automobile dealers, affixed to new cars, and included in new car advertising. Energy efficiency ratings have been calculated for air conditioners and other major household appliances, and manufacturers have been required to display this information prominently on the appliances. Pamphlets have been published offering advice on how to reinsulate homes, how to drive an automobile in an energy-saving manner, and how to save energy in many other ways. State Energy Extension Services have been funded, and each has provided its own set of information programs. A computerized energy audit system ("Project Conserve") was developed to provide accurate information on the energy-saving measures that should prove cost-effective for individual homes. And in the Residential Conservation Service, detailed energy audit information was included in a package of services designed to provide expert advice, financial assistance, ease of purchase and payment, and quality control for major household weatherization activities.

While some of those programs have been discontinued and others are in the process of change, all share a common implicit rationale that complete information is necessary to make rational decisions in the energy market and that, because of new developments in technology and continuous changes in energy prices, the information that individuals and households had previously used had become obsolete. Without programs to provide accurate information on the new environment, people would make decisions that would not further their own or the nation's interests. This rationale derives from a view of energy users as rational actors motivated to minimize energy-related costs and maximize income available after energy needs have been met. In short, the creators of government information programs have usually assumed that energy users act as investors.

Federal energy information programs have also proceeded from implicit assumptions about the way information works—and those assumptions are fundamentally wrong. The programs tend to be constructed as if people presented with accurate estimates of, for example, thermal performance of a variety of furnaces, would use this information in purchasing decisions. Even when people are acting as investors, however, this is not the case: information that reaches a person's eyes or ears is not necessarily noticed, understood, assimilated, or used. For information to be effective in a decision process, making it available is not enough.

The Roles of Government

Although the need to attract attention, offer clear and compelling presentations, and motivate audiences to use information is well understood by communications professionals, government officials have generally been unwilling to build energy information programs on principles of effective communication. This unwillingness is partly based on a fear of manipulation by government and on the notion that government should be responsive to the public, not the other way around. But there are some inconsistencies in the view of the proper government role. For example, it is considered appropriate in the United States for government to offer strong financial incentives such as low-interest loans, tax credits, and excise taxes to influence citizens and organizations to conform to the intent of public policy. It is usually considered proper to proscribe behavior with regulations, and to attach civil or criminal penalties. It is also considered appropriate for public officials, notably the president and cabinet members, to use access to mass media in attempts to persuade citizens to support their policies—or even their policy proposals. And it is considered appropriate for government agencies to use the media to argue against drug use and to promote good nutrition and other public welfare measures, although only if media access is available at no cost. It has usually been considered inappropriate, however, for the U.S. government to use paid advertising to persuade people to do what public policy implies or requires. And government agencies rarely employ communications professionals to design informational materials for maximum impact. Government agencies are in a bind: they can aid public policy by providing citizens with information, but they are seriously constrained as to methods. The guiding philosophy has been to make information passive, leaving energy users mainly responsible for searching out, selecting, and interpreting available information.

The roots of this situation are a fascinating part of the human dimension of energy. Clearly, political pressure from affected interests as well as legitimate fears of manipulation have a role in the design of particular informational programs. This is why some energy information programs have been pursued more aggressively than others and why the importance given energy information, and the form it takes, changes with national administrations. In particular, the Carter and Reagan administrations differed in their views of the proper role of the federal government in energy information.

While U.S. policy may change with administrations, the range of approaches taken in the United States is rather narrow. This can be easily seen by comparison with Canadian government information programs. Figures 4, 5, and 6 are reproductions of full-page newspaper advertisements purchased by the Canadian government's Ministry of Energy, Mines, and Resources in the late 1970s. In several ways, these informational efforts do things that U.S. government energy information programs have never done. They are, first of all, paid media advertising. Second, they are clearly promotional in tone. And third, they attack, in fairly direct ways, the interests of energy producers and automobile companies.

Edward Wall of Saint John, N.B., isn't happy about his fuel bills, but he hasn't added insulation.   *Ted Reinders of Saint John, N.B., added insulation to his attic and walls. Now he's saving energy and money.*

**Last winter, this
New Brunswick house
cost Edward Wall
$725 to heat.**

**Last winter, <u>this</u>
New Brunswick house
cost Ted Reinders
only $478 to heat.**

Why the difference? <u>Insulation.</u>
These days it's one of the best
investments you can make.

Keeping the heat in helps to conserve Canada's
dwindling energy reserves. And cuts your fuel bills.
    Add insulation, caulking, weather-stripping and
storms. Have your furnace tuned for peak efficiency.
Keep a light hand on the thermostat.
    You'll save money for yourself, and help Canada
to conserve energy and fight inflation. No wonder
insulation is one of the best investments you can
make. Remember, too, that many insulation products
are now exempt from Federal sales tax.

These two free books
show you how to save
energy and
money. Mail
the coupon
today.

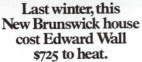

Please send me a free copy of:

"KEEPING THE HEAT IN", ☐

"THE BILLPAYER'S GUIDE TO FURNACE SERVICING". ☐
Check the appropriate box

NAME _____
(PLEASE PRINT)

ADDRESS _____

PROVINCE _____ POSTAL CODE _____

Mail coupon to **FREE BOOKS**, Box 900,
Westmount Postal Station, Montreal H3Z 2V1

If you're not part of the solution, you're part of the problem.

**Energy, Mines and
Resources Canada**
Office of Energy Conservation
Hon. Alastair Gillespie
Minister

**Énergie, Mines et
Ressources Canada**
Bureau de la conservation de l'énergie
L'Hon. Alastair Gillespie
Ministre

Fig. 4.   Canadian government advertisement for insulation (#1)

SOURCE: Energy, Mines and Resources Canada

    The difference between the Canadian and U.S. approaches appears to
lie partly in the extent to which energy conservation is defined as a public
interest in the two countries. This, in turn, relates to dominant conceptions
of energy, as described in Chapter 2. To the extent that energy is defined
as an ordinary commodity, informational functions are likely to be left to
producers of fuels and manufacturers of energy-using equipment, possibly
with some regulation to guard against fraud. If, however, energy is seen
as an ecological resource, a necessity, or a strategic material, a public
interest in controlling energy use is implied, and more aggressive governmental programs are justified. So far, these latter views of energy and the
interests they serve have not been sufficiently influential in U.S. politics

Josephine Webb of Hamilton now has six inches of insulation in her attic. She's saving energy and money.

Mrs. Josephine Webb of Hamilton:

"After insulating, I saved 178 gallons of oil. Better that the money be in my pocket than the oil company's."

Adding insulation is one of the best investments you can make.

Keeping the heat in helps to conserve Canada's dwindling energy reserves. And cuts your fuel bills.

Add insulation, caulking, weather-stripping and storms. Have your furnace tuned for peak efficiency. Keep a light hand on the thermostat.

You'll save money for yourself, and help Canada to conserve energy and fight inflation. No wonder insulation is one of the best investments you can make. Remember, too, that many insulation products are now exempt from Federal sales tax.

These two free books show you how to save energy and money. Mail the coupon today.

the billpayer's guide to furnace servicing

keeping the heat in
HOW TO RE-INSULATE YOUR HOME TO SAVE ENERGY AND MONEY (AND BE MORE COMFORTABLE TOO)

Please send me a free copy of:

"KEEPING THE HEAT IN", ☐

"THE BILLPAYER'S GUIDE TO FURNACE SERVICING". ☐
Check the appropriate box

NAME _____
(PLEASE PRINT)

ADDRESS _____

PROVINCE _____ POSTAL CODE _____

Mail coupon to **FREE BOOKS**, Box 900, Westmount Postal Station, Montreal H3Z 2V1

If you're not part of the solution, you're part of the problem.

Energy, Mines and Resources Canada
Office of Energy Conservation
Hon. Alastair Gillespie
Minister

Énergie, Mines et Ressources Canada
Bureau de la conservation de l'énergie
L'Hon. Alastair Gillespie
Ministre

Fig. 5. Canadian government advertisement for insulation (#2)

SOURCE: Energy, Mines and Resources Canada

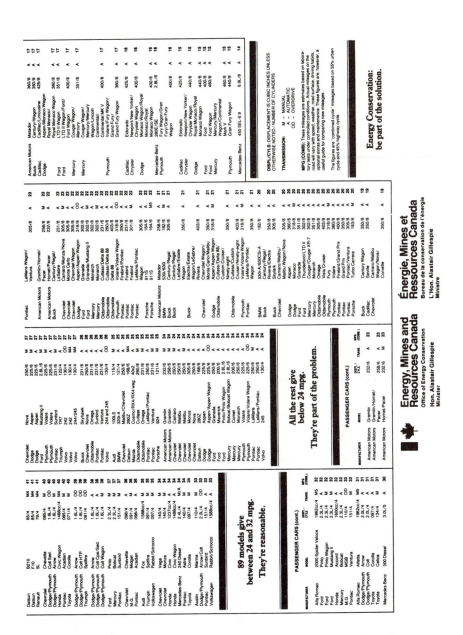

Fig. 6. **Full-page Canadian government newspaper advertisement for energy-efficient automobile purchases.**

SOURCE: Energy, Mines and Resources Canada

to overcome resistance, so there are no ads containing ideas like those supported by the Canadian government.

Unlike many energy information efforts in the United States, the Canadian messages are characterized by vivid pictorial and graphic illustrations, appeals to a variety of motives (solving a national problem, saving money, preventing loss of personal funds to the oil companies, and so forth), and attention to particular behaviors likely to follow reading the ad. These features are likely to increase the effectiveness of the communications. Since no clear rationale has ever been offered for keeping government communications dull and narrowly focused, it may be politically easier to change the format of information than the content or medium of presentation. Indeed, after years of apparent inattention, the Department of Energy has begun to experiment with graphics to attract attention to fuel economy stickers on automobiles (Figure 7).

Still, energy information programs in the United States continue to operate under serious limitations. The society's decision on whether to remove some of the present restrictions on information programs will depend in part on decisions about whether there is a public interest in increasing energy efficiency. It is important to recognize, however, that information programs that fail to present information in the most effective way, for whatever reason, will fall short of the hopes held out for them: information made available is not the same as information used.

The Value of Information Programs

For analytic purposes, it makes sense to judge information programs against the criterion of rational decision making that underlies them. An effective energy program would be one that leads energy users to take actions that minimize the total cost of the energy-related services they purchase—that is, to do what economically rational people theoretically do with full information.[11] Of course, expectations of effectiveness must also reflect resources devoted to a program.

Our discussion clearly suggests that judged against this criterion, existing and past energy information programs are ineffective. However, relatively few energy information programs have been formally evaluated, and the available evaluations have been much less systematic than is desirable for drawing firm conclusions. True experimental studies of energy information programs are rare, so the evaluation research is plagued by problems of inference: because program participants are volunteers, they differ from nonparticipants, and it is difficult to tell how much of the change observed among participants might have occurred without a program. In addition, many evaluations rest on questionable assumptions about the validity of respondents' self-reports of their behavior; about the actual behavior implied by survey responses, such as checking a box marked "insulated attic;"

CURRENT LABEL

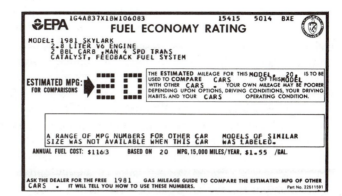

ALTERNATIVE LABEL TESTED

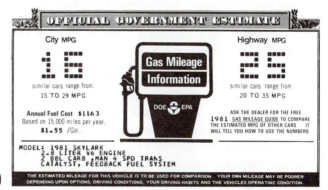

ALTERNATIVE LABEL TESTED

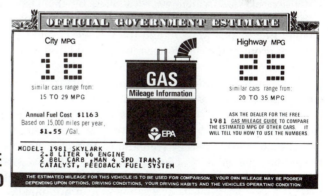

Fig. 7. Fuel economy label designs

SOURCE: Pirkey et. al., 1982

about the energy implications of the behavior if it did occur; and so on. Finally, because most of the evaluation studies were commissioned by organizations responsible for the programs being evaluated, who have a stake in the outcomes of the research, the assumptions, methodologies, and conclusions may have been subject to political pressure, with unknown effects. For example, the Department of Energy commissioned an evaluation of a 1979 program in which it distributed water flow limiters and conservation information to all households in New England (Booz, Allen, and Hamilton, 1980). It found an annual savings of $26 in fuel for each dollar spent by the program. But a review by the General Accounting Office (GAO) (1981) criticized the evaluation for making assumptions that may have vastly inflated the estimates of energy and money saved. The Department of Energy subsequently reestimated energy savings to account for one of the GAO's major criticisms; this reestimation reduced the estimated savings by one-third. The remaining GAO criticisms did not receive response, but a defensible case can be made that even the revised estimate is several times too high. Although on the surface the program seems to have been cost effective, this conclusion cannot be drawn from the evaluation with much certainty.

Because of the self-interest of most of these conducting evaluation studies, it would be naive to take their conclusions at face value. But what of the general issue: how valuable is energy information? A careful analysis of the methodologies and data convinces us that some energy information programs have been effective in terms of producing economically justified energy savings by consumers. These programs probably include the Energy Extension Service, which was evaluated by the Department of Energy (1980), the automobile fuel economy information program of the Department of Energy (see McNutt and Rucker, 1981), and a few of the programs aimed at promoting major home retrofits (see Hirst, Berry, and Soderstrom, 1981).

Other lines of evidence also suggest that under optimum conditions, energy information programs might be important. It is obvious, for example, that automobile manufacturers know that fuel economy sometimes sells cars: their advertising has often said much more about miles-per-gallon than is mandated by regulations. In fact, it has been estimated that the automotive industry spent upwards of $100 million on advertising fuel economy in 1979 (Pirkey, McNutt, Hemphill, and Dulla, 1982). While fuel economy has been less of a concern of automobile advertisers since then, in 1982 it was still one of the major foci (Pirkey, 1982), reflecting the concerns expressed by purchasers of new cars. At the same time, because manufacturers are not trusted as a source of fuel economy information (Pirkey, 1982), information from a standardized fuel efficiency test has an important function—if it is effectively presented.

Overall, most information programs seem to do less than their creators hope. This occurs both because information that is made available does not reach all of the intended audience and because the information that does reach the audience often remains unused. But because the implementation of information campaigns has been severely restricted it would be a mistake to conclude that energy information *cannot* be effective. As Ester and Winett (1982) have shown, the frequent finding of ineffectiveness reflects the fact that the interventions evaluated are usually of poor design.

Problems and Opportunities

In Chapter 3, we identified several conditions that partly explain why information does not always lead people to take all actions that are in their own economic interest. Some of these conditions are outside the control of any information program: lack of funds; inability to invest in property that is only rented or leased; unavailability of desired energy-saving products; and the fact that many households, because of their own or someone else's previous decisions, own, live in, or operate energy-inefficient houses, apartments, automobiles, furnaces, or appliances that are prohibitively expensive to replace. But there are still other problems and opportunities for information programs.

Government programs have emphasized developing energy information that is accurate and reliable (in a scientific sense) and disseminating the information to reach many people. These emphases are necessary, but there are other problems that must also be addressed if an information program is to be effective:

Scarcity of Attention.   Energy users sometimes act as problem avoiders. For many people, processing complex energy information presents a formidable task. Information is less likely to be used if it arrives when a householder is busy with more pressing issues; if it is too detailed for the energy user's needs; if it is not easily related to the recipient's daily concerns; or if it is presented in a manner that fails to attract attention.

Diversity of Energy Users' Needs.   Because households use different fuels in different proportions for different purposes, because they live in different kinds of buildings, because they have different travel needs, and because of variations in income and housing tenure, general information includes much that is irrelevant for any given energy user. This may distract from relevant information or make the provider of the information seem unreliable or useless. But the problem of diversity is not generic. It is most serious for information about such technologies as solar heating and home insulation, for which the energy savings vary greatly with the building, the climate, the manner of installation, and users' behavior. It is less serious for mass-produced products like automobiles and air conditioners, because, even though these products are used differently by different people, manufacture is standardized. The problem of diversity is least serious when neither the technology nor user behavior varies greatly, as with refrigerators and freezers.

Invisibility of Energy. Because most people have little experience with the details of the operation of furnaces, water heaters, automobile engines, and so forth, they may not pay attention to information on how to improve their operation or find more efficient equipment, or they may find such technical information incomprehensible. Invisibility is also a problem because many of the remedies recommended by information programs are themselves invisible, so it is difficult for an energy user to know whether or not the remedy has been effective.

Understandable Skepticism About Energy Information. A government agency may be convinced that its information is accurate and reliable and that its estimates of savings are reasonable, but energy users have good reasons to be skeptical. This problem is particularly difficult because a solution seems to require information providers to recognize that no matter how carefully and scientifically they gather information, people will need to consult other sources before they can develop trust.

Informational programs can address these problems. We first discuss two topics that have been addressed by much careful research: the problems of attracting the attention of energy users and of making energy savings visible. Then we turn to home energy audit programs to show how this knowledge can be used.

Attracting Attention

Because information is effective only if it attracts attention, telling people how to save energy can be seen as an advertising problem. But there are important differences between offering energy information and selling soap or toothpaste. The concept of "social marketing" (Kotler and Zaltman, 1971) is useful for pointing out these differences.

Social marketing has been defined as the attempt "to increase the acceptability of a social idea or practice in a target group(s)" (Kotler, 1975). Generally, these ideas or practices have benefits beyond the individual who adopts them. Thus, offering information on energy efficiency may be a form of social marketing. To consider it as such, however, is to question the idea of passive distribution of information that underlies much of the U.S. government's effort to offer energy information. "Marketing" implies a much more active process. This points to a major difference between selling soap and providing energy information: to the extent that energy information programs are publicly supported or mandated, they are the outcome of a political process that determines a public interest is being served. This process not only creates information programs, but influences their content and methods of operation.

The political nature of energy information programs creates problems for those who run them. Bloom and Novelli (1981) point out that social marketers often face pressure against treating different segments of their audiences differently because this may be seen as unfair. Thus, it may be politically difficult to use the techniques of market segmentation that have been developed by the advertising industry. In addition, politics often limits the choice of media and messages for energy campaigns. An example is the prohibition against government purchase of advertising space or media time for energy information.

Social marketers are subject to political pressure in a way corporate advertisers rarely are. As a result, public information programs in the United States have been prevented from using the most effective communication techniques. However, private organizations, such as utility companies, can carry out programs of social marketing of energy conservation without public debate. While their actions would not be so restricted by political control as those of government, they are restricted by private control—the public interest is not usually a prime focus of privately run energy information programs.

Another difference between energy information and ordinary advertising is organizational. Information providers in government cannot influence product design or pricing the way the marketing department of an automobile or of an appliance manufacturing company can. Yet the design and pricing of a device such as a miles-per-gallon monitor for new cars may be an important influence on its acceptability.

Energy information is also unlike soap or toothpaste because of the nature of the "products." Like other objects of social marketing (Bloom and Novelli, 1981), energy efficiency is not well understood by the public. One reason for this is the invisibility of energy flows. As a result, it is difficult or impossible to design simple messages or "product concepts" that can, through repetition, make an impression on an audience. Energy information is more like an educational program than an ordinary advertising problem in that its messages are complex. It is also often essential in social marketing to "sell" ideas to those least likely or able to "buy," for example, antismoking messages to confirmed smokers or energy efficiency to low-income households.

Nevertheless, available knowledge could be better used in the design of energy information programs, despite all the existing constraints. If information for energy users is to be effective, it must be presented to attract and hold attention. To do this, the information must be presented in an inviting format, it must be made easily understandable, it must be specific to the desired end, it should be timed so that action by the consumer is convenient and possible soon after the information is received, and it should be distributed by sources that command the energy user's attention (Ester and Winett, 1982; Stern and Gardner, 1981). Energy information programs have not generally followed these principles: energy information is often available only in monochromatic, small-print pamphlets distributed upon request by the Government Printing Office. Although much of the information is valuable, the format is neither eye-catching, simple, nor convenient, and the distribution system is unknown to most people. Even the best-recognized energy information program, the fuel economy labels on new cars, has been presented for nine years in a dull format (see Figure 7, above). Other communication-based programs have been too general to be effective—for example, a billboard that shows the national gas tank reading nearly empty, with the legend, "Don't be fuelish."

Dozens of controlled experiments have been conducted using various sorts of communications to get people to take actions that reduce use of energy or other resources. Reviewers have generally concluded (Ester and Winett, 1982; Shippee, 1980; Winett and Neale, 1979) that the experiments have produced very little energy savings. But like many government programs, most of these studies have failed to use the most effective available communication techniques (Ester and Winett, 1982). The evidence from this group of studies supports the knowledge of communication theory, emphasizing the effectiveness of messages that are specific, repeated, and very close in time to the desired behavior (Ester and Winett, 1982).[12]

Not all information programs have been ignorant of communications principles. There has been some noticeable success in presenting information in simple and understandable form. Some programs have distilled energy-efficiency information into simple numbers—miles-per-gallon, energy efficiency ratios for air conditioners, and payback estimates.[13] Other programs have created a simple category, such as when utilities have certified certain buildings as energy efficient, thus providing simple and valuable information to prospective purchasers or renters (Stern et al., 1981).

The miles-per-gallon number has proven especially useful, judging from the frequent emphasis on it in automobile and motor oil advertising and in casual discussions among motorists. Miles-per-gallon is an especially meaningful unit for energy users, being a ratio of two measurements that are meaningful and generally used (Kempton and Montgomery, 1982). Energy-efficiency numbers for appliances and payback-period estimates are probably a step in the right direction, since they simplify information processing. However, because they do not build on familiar and intuitively meaningful concepts, they are likely to be less potent than the miles-per-gallon concept.

Making Energy Savings Visible

Many of the most effective energy-saving devices are hard to see. Moreover, the energy they save is invisible and difficult for most energy users to monitor with any accuracy. Other effects of energy savings, such as increased comfort or a small increase in disposable income, may also be difficult to perceive. As a result, people who successfully save energy do not necessarily know it and may even conclude that their efforts are useless. For example, consider a homeowner whose savings from adding insulation or a new furnace are masked by changes in weather or energy prices. If cooling or heating costs are higher *after* the investment than before, it is discouraging.

For these reasons, energy information programs should benefit from making energy savings more visible to those who achieve them. Such programs would gain credibility and the benefit of positive word-of-mouth advertising. When people learn the difference between effective and ineffective energy-saving actions they become better able to respond to emergency conditions that call for rapid changes in patterns of energy use.

Field experiments have explored various ways to provide energy users with an accurate account of their energy use. One method involves the installation of devices that monitor fuel or electricity use and provide digital readouts in cents-per-day, miles-per-gallon, or other useful units. In another method, people read gas or electric meters and provide householders with energy-use information on a daily or weekly basis, again using any of various simplifying measurements. Research has examined the effects of teaching people to read their own meters, sometimes combining this instruction with a commitment from the energy user to try to save a certain percentage of energy use. Still other research has examined variations on the presentation of information on utility bills, maintaining the regular monthly billing period, but providing information in a more useful form.

Many of these "feedback" procedures have proven effective in studies of residential energy use (Seligman, Becker, and Darley, 1981; Shippee, 1980; Winett and Neale, 1979). The degree of effectiveness depends on several factors. First, feedback must be credible—that is, related to behavior. For feedback about energy used for home heating and cooling, this usually means using weather-corrected units. But if the units being used are not understood, the information may be discredited (Winett and Neale, 1979).

Second, feedback is more effective when energy users have made a commitment to conserve energy (Seligman et al., 1981) or have set themselves quantitative goals for saving energy. In one study, participants in a feedback experiment who committed themselves to reducing energy consumption 20 percent used 12 percent less energy than people who were trying to save only 2 percent (Becker, 1978). But participants did not make additional savings simply by setting a more difficult goal. In comparison groups that did not receive feedback, setting a more difficult goal produced only a 5 percent relative savings of energy. Feedback is more likely to be effective if given as part of a program in which the energy user is an active participant rather than simply being a passive recipient—even when information is offered about how to interpret the feedback.

Third, there is evidence to suggest that more frequent feedback is more effective. Frequency may be important mainly for individuals with relatively little commitment or when the energy-saving behaviors are difficult to maintain (Seligman et al., 1981). It may also be that frequency per se is less important than the fact that the feedback follows immediately upon some action that has been tried for saving energy (Shippee, 1980). If this is the case, what is important is that the feedback be meaningful for behavior—someone might learn as much by occasionally monitoring a utility meter or more sophisticated device as from a program of frequent feedback. In addition, frequent feedback that simply appears regularly and is unrelated to behavior may soon be seen as irrelevant and be disregarded.

Fourth, feedback is more effective when the relevant energy costs are a large portion of the household budget (Winkler and Winett, 1982). While this finding demonstrates the importance of energy prices, it also underscores a point we made in Chapter 3 about energy invisibility—that economic forces are not enough to bring about the appropriate adjustments in household energy use. Rising prices or falling incomes provide motivation and bring energy into awareness, but motivation alone produces less saving than motivation combined with a technique to make energy more visible. As Winkler and Winett (1982) found, feedback typically had no effect when energy costs were no more than 2 percent of household budgets; by contrast, when energy costs were more than 5 percent of the household budget, feedback produced energy savings of around 15 percent beyond the effect of cost alone. A reasonable interpretation is that energy costs motivate and that feedback provides essential knowledge about how well efforts to save energy and money are succeeding. This is consistent with the notion that energy invisibility has involved a loss of knowledge among residential energy users that has important implications for energy use.[14]

It is important to note that most of the experimental research on feedback provided information by highly labor-intensive methods, often involving daily meter readings and individualized communications for participants in the research. Such procedures are not necessary; feedback can be automated at fairly low cost with display meters. Display meters, installed in a home, would solve the problem of lack of frequent and meaningful feedback by making feedback available whenever needed. But there has been too little research to determine what format for feedback would be the most credible and useful. The question is not trivial. In the first study of residential display meters, the feedback had no effect (Becker et al., 1979)—possibly because feedback was given (in cents-per-hour) at two-second intervals. This method showed major changes in energy use only when appliances were shut off entirely.

Useful feedback might also be delivered to households through an improved utility billing system. Research on this issue, while limited, shows that it has some promise as well as some problems (Russo, 1977).[15] A few studies have also examined the use of energy consumption feedback in motor vehicles. They are somewhat promising, showing small fuel savings in the 5 to 10 percent range from the installation of devices to monitor fuel use on cars and trucks (Stern and Gardner, 1981; Reichel and Geller, 1981). It is reasonable to expect that the effectiveness of feedback on fuel used in vehicles will depend on some of the same variables as for home energy use: credibility of the feedback; commitment and the setting of goals; frequency; and the importance of fuel costs in the user's budget.

Feedback systems for making energy visible have a potential advantage over verbal information and advice in that they can be made credible independently of the issue of trust in an information source. While a utility's motives may be mistrusted when it sends advice on how to save energy, its meter readings are usually accepted as definitive. Therefore, feedback systems may be important independently of issues of trust.

HOME ENERGY AUDITS

Because houses and apartments vary so much in their structural characteristics and because their energy requirements also depend on climate, the cost of fuel used, the stock of appliances in the home, and the behavior of their occupants, most recommendations for energy-saving activity in "all homes" are likely to be inappropriate for many households. Technical experts in residential energy conservation have long recognized this and have developed the concept of a home energy audit specifically to handle the problem of diversity—to give householders expert advice suited to their individual situations.

Over almost a decade, various programs to provide individualized energy information through home energy audits have existed in the United States. Some of them have been based on short building surveys, filled out by residents and analyzed by computer to give recommendations for action. These "Class B" energy audits typically give the householder information on the estimated costs of each suggested action (if installed by a contractor or if done with the resident's own labor) and some simple measure of expected return on investment—either estimated dollar savings per year or an estimate of the number of years it would take the investment to pay back its cost at current energy prices. Class B audits are an inexpensive way to give householders individualized recommendations, but they are based on a highly simplified analysis of the building and thus do not give the most accurate information for a given home.

A "Class A" energy audit involves a visit by a trained energy auditor to the home and an inspection of the residence; it frequently results in on-the-spot recommendations to the occupants. It is more expensive, but the expense may be worthwhile because more accurate information results. Class A audits have been a required part of programs under the Residential Conservation Service, a national program mandated in the Energy Conservation Policy Act of 1978. The best available documentation of a rationale for home energy audits is in the regulations for the Residential Conservation Service (*Federal Register,* Nov. 7, 1979).[16]

The original RCS regulations emphasized the need of household energy users for useful, reliable, and accurate information. More concretely, the regulations detailed recommended conservation practices by heating fuel and by climatic zone—down to the county level. They described standards for materials to be used in recommended work under the program; and they presented formulas for calculating cost-effectiveness before recommending conservation measures. These efforts were made to ensure that auditors' recommendations would be based on the most reliable and accurate available information. The regulations also emphasized that energy auditors be trained, though procedures for training were not spelled out. In addition, RCS regulations required that recommendations be delivered to householders in person. The regulations also emphasize household economics: auditors are required to recommend only those actions that are justifiable in terms of payback time, and payback is a prominent feature of the information energy auditors provide. This presumably makes the information useful.[17]

The Effectiveness of Home Energy Audits

Like other energy programs, RCS apparently assumes energy users are primarily rational actors who wish to minimize energy-related costs over an extended period and who need full information to make the most effective decisions. The attention to the use of the payback concept shows that RCS also recognizes that full information, depending on how it is presented, may be more or less useful. However, the evidence shows that the useful, reliable, and accurate information in most RCS and related programs is not necessarily effective information.

The information in energy audits is often ineffective because it reaches so few people. Most utilities that run energy audit programs report participation rates of less than 5 percent of eligible customers (Hirst et al., 1981; Rosenberg, 1980), despite their publicity efforts. And this does not mean that few people need the information; actually, the people in households that request energy audits are generally more educated, have higher incomes, and are more interested in energy issues than the general public (e.g., Hirst et al., 1981). This suggests that the information is going to people who are likely to be better informed in the first place. It also means that the poor and less educated, who have a strong need for information and who tend to live in energy-inefficient homes, are not reached by these information programs. Eric Hirst and his colleagues (1981) identified this as the most obvious shortcoming among the twenty-seven utility-run programs they studied.

Furthermore, those who get the detailed information from energy audits do not necessarily act on it. One review (Rosenberg, 1980) concluded that only 20 to 30 percent of households followed their participation in programs that combined energy audits and low-interest financing by taking the recommended action. And another review (Hirst et al., 1981) concluded that the level of energy-saving activity among participants in audit programs represents only a modest increase over the level of action by nonparticipants. Thus, energy audit programs have often fallen short on two counts: reaching their clienteles, and converting audits into action.

Reaching the Clientele

The evidence suggests that householders respond to audit programs partly as a function of the method used to publicize the programs. Rosenberg (1980) reports participation rates of 1 to 3 percent when publicity is by enclosures with utility bills; 3 to 6 percent with direct mail: 5 to 7 percent with a combination of direct mail and media advertising; and higher participation when unpaid media news coverage is made available. Stern, Black, and Elworth (1981) have pointed out that such publicity techniques are probably most effective with middle-class homeowners, while for a low-income clientele, word-of-mouth campaigns relying on community groups, tenants' associations, church groups, and so forth might be more effective. They suggest that large organizations sponsoring audit programs may be more effective in getting attention if they join forces with local groups that have personal contacts with the intended clientele.

Aggressively pursued information programs can be effective with low-income groups when the programs emphasize energy-saving techniques appropriate to their clients. One study (Winett, Love, and Kidd, 1982) tested a program in which energy experts visited residents of centrally air-conditioned low-income housing units in Virginia and demonstrated methods for minimizing use of electricity for water heating and cooling. The program lent each participating household a window fan and provided instruction on where and when to use it, on the use of natural ventilation for cooling, and on energy-efficient operation of the water heater. Savings ranged from 9 percent in cooler summer periods to 24 percent when the weather was hotter.

A few generalizations about publicity for energy information campaigns have emerged from a review of media campaigns by Koster (1981): programs were more successful when they offered a premium, such as a water-flow restrictor or free information booklets, and when they combined appeals to consumer responsibility with suggestions for specific actions. These recommendations are consistent with Hutton's (1982) findings on energy advertising campaigns and with the importance of commitment and personal control. Koster also found that scare tactics and appeals to patriotism were much less effective than specific, useful information. In addition, Koster's report offers numerous suggestions about operational details involved in using the mass media for energy information campaigns.

Publicity for energy audit programs can also take advantage of the diversity in energy users' needs and their motives for saving energy. For some people, saving energy is a matter of financial necessity, while for others it may mean eliminating cold drafts, conserving scarce landlord-supplied heat, or attaining the satisfaction that comes from upholding personal values of thrift of environmental preservation. Announcements of energy audits, as well as communications from program personnel during the audits themselves, should emphasize those outcomes of energy savings that are important for the sort of household they are addressing. Communications should be different for homeowners and renters: for people who are or are not concerned about national energy problems; and for people who differ in other identifiable ways. It is also useful to have several sources distribute information in each area. In this way, more people will receive information from at least one source they pay attention to and trust. If a program is publicized through local organizations, the staffs of those organizations may be in the best position to know what their members and associates want from an energy program. This knowledge can do more than help with publicity. It may also direct a program's emphasis to particular kinds of energy-saving actions—do-it-yourself versus contracted improvements, for example.

Publicity cannot overcome all the reasons for nonresponse to audit programs. Some programs emphasize activities that cost too much for certain households or are irrelevant for renters. And some program sponsors may not be motivated to try to reach all the clientele, as when an electric or gas utility company is responsible for providing energy-saving information to households that heat with oil.

From Audit to Action

Assuming that an energy auditor gets in the door, the effectiveness of an audit still depends on much more than having reliable, understandable information to offer. The energy audit process should be treated as one of interpersonal communication rather than one of machinelike information transfer, because: (1) people do not assimilate all information equally; (2) the credibility and trustworthiness of information sources are paramount issues; (3) what people do with information may depend greatly upon the communicator's social skills; and (4) it is necessary to communicate information that is unfamiliar and hard for many people to visualize.

An emphasis on the interpersonal aspect of an energy audit immediately underscores the importance of the RCS regulation, recently eliminated, that mandates that information be delivered in person. While important, however, personal contact is not enough to make the energy audit an effective communication. Since people have learned to be skeptical of energy information, credibility is of critical importance—and an essential component of credibility is trustworthiness. Because of the low level of trust in some energy institutions (Farhar, Unseld, Vories, and Crews, 1980), energy auditors will often fail to elicit trust even with personal contact. Energy auditors are likely to be more effective if they are not employees or agents of anyone who can profit from their recommendations.

Getting Messages Noticed. An auditor is also much more effective if he or she presents information in ways that attract attention and are memorable (Borgida and Nisbett, 1977; Hamill, Wilson, and Nisbett, 1980; Nisbett et al., 1976; Taylor and Thompson, 1982). Statistical data summaries and impersonal informational sources are less salient than face-to-face interactions and detailed case studies. As with the story of the prospective Volvo buyer (Nisbett et al., 1976), impersonal data summaries, however accurate and reliable, have less impact than more concrete information even when the more vivid and personal information is less representative.

Thus, energy audits are more effective if auditors are trained in communication skills. They should be trained to present information not only completely and accurately, but also in the most vivid and salient manner possible. This may include, where feasible, reports of energy savings achieved in similar homes in the same climate zone. Even better are accounts of homes in the same neighborhood—and reports from the homes of friends the householder has visited would be still more dramatic.

Probably because of its salience and vividness, personalized information has more impact on individuals than statistical data, even when the latter are much more reliable and accurate (Nisbett et al., 1976; Taylor and Thompson, 1982). Intuitive knowledge of this phenomenon is sometimes applied by unscrupulous salespeople to mislead, but it can also be used to emphasize a statistically valid point. Thus, as Yates and Aronson (1983) suggest, after describing the average cost-benefit ratios associated with various conservation practices, an energy auditor might describe the experience of a "superconserver" as well. Superconservers are families who save more energy and money than the average. The auditor might say something such as the following: "... of course, these estimates are averages; your savings might be greater or less than average. But, I'd like to tell you about one local success story so you can get a sense of just how effective these practices can be. The 'Smiths,' who live in this town, installed weatherstripping. They save about [the average savings of this superconserving family] every month." Identifying superconservers provides tangible and dramatic evidence.

This approach may seem a bit like high-pressure salesmanship and therefore offensive. Such a reaction would be particularly likely and appropriate if the audit program were sponsored by an organization that is mistrusted. Thus, the acceptability of such an aggressive approach depends on the earned trust of the auditor or the organization represented. One way to gain trust would be to encourage word-of-mouth communication about the program. For example, people who receive audits could meet or talk to "superconservers" and other clients of the program.

The importance of salience and vividness, personalized information, and trust suggest that energy auditors can be more effective if they work in a neighborhood and if they know people in the community. The idea of vividness also implies that a "hands-on" approach involving clear demonstrations would be greatly superior to a procedure in which the householder is passive and receives only a summary of recommendations. Telling people that they are losing a certain percentage of home heat through the cracks around their windows is not as compelling as setting up a situation that demonstrates the heat loss with an infrared scanner or shows the smoke from a smoke stick pouring out under doors and over window sills. Smoke sticks emit a fine powder that flows with the air currents in a room. They are inexpensive as well as quick and easy to use; they make energy losses highly visible and provide a graphic illustration of the need for caulking and weatherstripping.

The "House Doctor" concept, as developed at Princeton University, is an excellent example of an energy audit system that has translated some of these principles into action. Energy auditors were trained to encourage the householder to go on the audit with them around the home. The house doctors explain the audit, demonstrate with smoke sticks, and invite active participation on the client's part.

It is relatively ineffective for energy auditors to simply leave the householder with a computer summary of the potential savings associated with various retrofits. It is more effective for the auditor to discuss all recommendations with the customer. Moreover, if an auditor uses a copy of the household's most recent utility bill to illustrate a point, it is personal and more effective than talking in the abstract.

"Framing" Messages.   Recent research has shown that information is more or less effective depending on the way it is "framed" (Tversky and Kahneman, 1981). An illustration of this is the evidence that while most people might drive across town to save $5 on a $15 item, few would drive across town to save $5 on a $125 item. Rational actors would be equally likely to make the effort under both conditions, but for real people, the situations are apparently very different.[18] Another example of the importance of framing a situation is that the amount of joy experienced when someone wins $100 is not equal to the consternation suffered when that same person loses $100. Related to this fact are research results indicating that individuals are more willing to take a risk to avoid or minimize a *loss* than they are if the purpose of the risk is to *increase* their fortunes. Thus, people who fear a loss will probably be more open to innovation. This finding implies that information emphasizing payback and return on investment, because it stresses the amount of money and energy which can be *saved* or gained through conservation, is not optimally effective. Telling people how much they can save by investing money in conservation or alternative energy sources encourages people to define this as a "gain" or "win" situation. Once the situation is labelled as such, people express a reluctance to accept risks or depart from the status quo. Thus, this campaign strategy may actually be inadvertently discouraging people from changing their habits of energy use. But if auditors clearly framed energy conservation as the avoidance of loss—by showing residents how much they were losing every month by not investing in alternative energy sources and other conservation measures—one would expect a very different reaction. Once the loss becomes salient, people are encouraged to take more drastic action.

A recent study by Yates (1982) provides some direct support for this notion as it relates to energy. Homeowners were asked to evaluate some cost-benefit information about either a solar water heater or an insulating blanket for the water heater. The presentations were designed to focus either on potential savings if the investment was made or on continued losses if the investment was not made. The homeowners evaluated the worth of products on a number of dimensions and indicated their intent to install a device in the coming year. The findings generally confirm the contention that homeowners find energy-efficient technology more attractive when they consider the negative consequences of inaction.

Commitment and Choice.   The energy auditing process can also make use of what is known about the effects of commitment on action. We have already mentioned research that demonstrates that when people make a public commitment, set themselves an explicit goal, or voluntarily take a small action in accordance with their wishes to conserve energy, they are likely to go on to take more significant action. This can be applied in an energy audit if some small actions can be taken on the spot. Thus, an energy auditor may show a householder how to install a flow limiter in a shower or how to weatherstrip a window, using materials provided by the audit program. To gain the effects of behavioral commitment, it is important to encourage people to do some of the work themselves. Weatherstripping may be especially effective because it can be used to combine hands-on experience, behavioral commitment, and, using smoke sticks or similar devices, a vivid before-and-after demonstration of effectiveness.

Even with small improvements like using weatherstripping material and flow-limiters, an energy audit should give an energy user a choice among actions to take. Allowing a choice increases commitment to the chosen action and also increases the householder's sense of control over his or her situation. As we have noted, the sense of control has a great effect both on a person's sense of well-being (e.g., Langer and Rodin, 1976) and on the willingness to accept suggestions from outside authorities (Brehm and Brehm, 1981).

Choice has also been an important factor in the adoption of energy-conserving measures, such as the automatic day-night thermostats studied at Princeton University (Becker, Seligman, and Darley, 1979) and the gasoline-saving device the U.S. Army tried and rejected (Thomas et al., 1975). Allowing choice, as with the override mechanism on the automatic thermostat, serves a purpose even if the available choice is never actually made. Choice makes information more acceptable and increases commitment. Involving people in the decision-making process and allowing them more control over their destinies can make an energy audit more effective at the time and encourage further efforts at savings in the future.

To summarize, then, to be effective an energy audit has to do much more than provide accurate, reliable, and useful information. The information must be organized to make the auditors credible sources, and the auditors should supplement their accurate information with vivid, personalized examples, hands-on demonstrations, and a choice of activities involving householders that will save energy on the spot. Most of these changes can be made in the audit process easily and inexpensively simply by training the auditors in communication skills.

Credibility and Program Design

Other problems of energy information, especially the credibility issue, call for efforts beyond training auditors. They require attention to the design of the program. Some of the most attractive features of RCS are its inspection and contractor-listing features—two design elements that build credibility. Still, distrust of the source of energy audits may make it very difficult for audits to be effective, even if they are well publicized, offer accurate information, and make highly sophisticated presentations. Doubtful information sources—which probably include many of the utilities responsible for RCS audits—have serious credibility problems that must be overcome if their programs are to be effective.

There are several strategies for doing this. One is to achieve cooperation between a low-credibility information source (which may be able to offer a program essential funds and expertise) and another information source that is better trusted by the clientele. The latter will want to be sure its credibility is not damaged by the association with a doubtful source and will, therefore, make demands on the program. But if it comes to accept the audit program as credible, the action of this trusted source should help to communicate trustworthiness to its membership or clientele. This sort of partnership between a utility or other large sponsor and small community groups has been tried with success in Rochester, New York, and is being pursued elsewhere (Talbot and Morgan, 1981).

A promising variant of this approach involves building on the systems that already are most effective in providing individuals and households with information about energy-saving activities—informal social networks. A program can gain the considerable benefits of word-of-mouth publicity by using people who are part of local social groups, relying on neighborhood organizations to announce its existence or deliver its services, and doing its publicity or disseminating its services on a local or neighborhood basis. Some programs have used a "Tupperware party" approach, in which energy audit information or installation instructions are delivered to a group of householders meeting in the home of one of them (Olsen and Cluett, 1979; Fitchburg Office of the Planning Coordinator, 1980). This brings neighbors together in a way that can create a network in which people talk about saving energy. Such informal social networks provide highly credible information. They can be incorporated in program design and implementation alone or in combination with other techniques. It is in the interest of an effective program to make sure that the information entering the social networks in any way is accurate. This can be done, for example, by installing energy-monitoring instruments in selected homes for demonstration purposes.

Another strategy is to create a new organization to run an audit program. At first such organizations will lack public recognition and may need a lot of publicity, but they can get off to a good start with support from state officials, as they did in Massachusetts and Rhode Island (Rosenberg, 1980). After that, however, they must prove trustworthiness by their performance.

A third strategy is to organize a program so as to encourage and ensure good work: strict standards for materials, independent inspection of work resulting from the energy audits, independent conflict resolution mechanisms, and other consumer protection features are examples of useful procedures. Some procedures like these are incorporated in the regulations for RCS, and some of these have demonstrably increased the effectiveness of programs in encouraging energy-saving activity (Stern et al., 1981). Another procedure to encourage good work is the so-called Bradley Plan being tried in an innovative conservation experiment in New Jersey (Stern et al., 1981). In this plan, a private company offers free energy audits and free energy-saving home improvements to homeowners and earns a return on this investment through payments by the utility companies on the basis of actual energy savings. Although many problems can be anticipated in working out the practical details of this approach, the procedures give the energy service company clear incentives to provide good information and competent installations.

A program can increase its credibility if its energy auditors are well trained, if it recommends competent contractors, and if it is structured to reward auditors—and contractors, if they are involved—for accuracy and thoroughness. To produce a high level of participation may take time because distrust can only be removed by an extended record of trustworthiness. But because of the general distrust in energy information, proof of trustworthiness in action is the only way to have an effective energy information program.

A fourth way to encourage trustworthiness and build credibility for good programs is to get reliable information to energy users about which energy programs can be trusted. Energy programs may themselves take on the task of demonstrating their trustworthiness to their clients. If a householder can see clearly the effects of his or her actions—preferably in terms of money not lost, the comparison between actual energy costs and what they would have been without action—a program's effectiveness is vividly demonstrated. It may benefit an effective conservation program to monitor its clients' energy use before and after participation, and to feed this information back to the household in meaningful units of saving.

It is also possible to develop independent institutions to gather and disseminate information about energy programs. For mass-market products, including those that claim to save energy, a credible magazine like *Consumer Reports* serves this function. For information about energy programs and about the services of contractors who install energy-saving equipment, such a national approach is inappropriate because the services are not national. But it might be possible for similar institutions to be established at the local level. Like *Consumer Reports,* they would only be credible if they maintained high standards of objectivity and remained free of ties to interests other than those of consumers. It is not clear whether such information sources could be funded by subscribers or would need independent support. But experiments with independent information sources would be valuable because of the obvious need for credible sources of energy information for household energy users.

Beyond Credible Information

Improved energy audits, even if performed by credible experts, are not sufficient to make a program such as RCS optimally effective. There are other elements needed for an effective residential conservation program, and behavioral knowledge can be useful in designing some of them. In addition to publicity, important issues include assuring consumer protection, promoting trust, eliminating conflicts of interest, certifying contractors, assuring competition and consumer choice among service providers, and establishing effective relations among the organizations whose actions affect the program's success.[19]

Even with its organizational needs met and problems of conflict of interest overcome, a program such as RCS would not be enough to produce all the economically justifiable improvements in residential energy efficiency. For example, RCS has not been designed to meet the need for energy efficiency in rental housing, and it does not provide funds for households that lack sufficient money or credit for major energy saving expenses. But attention to the issues discussed here can improve the effectiveness of programs that rely on energy information among households who are able to take advantage of the information. Whether governments will develop or require more effective information programs depends on the definition of a public interest in energy efficiency, allocation of sufficient resources, and willingness to adopt some of the more aggressive approaches that make information more effective.

SUMMARY AND IMPLICATIONS

Five different views of the individual energy user contain important elements of truth. First, household users are investors; they are motivated to get the services that energy provides at the lowest net cost or to achieve the greatest possible return from investments in energy efficiency. Second, energy users are consumers; their energy-related activities have value to them in terms of comfort, esthetics, and similar goals—not just in financial terms. Third, individuals and households are members of social groups; they get information and ideas from their peers who, by communication and example, can exert considerable influence on energy-related behavior. Fourth, people seek to express their personal values in the things that they do—including things that use energy. Fifth, people try to avoid problems: they often persist in patterns of energy use that are not economically beneficial, but that could be changed only with effort; they tend to make decisions based on simple rules of thumb, confidence in someone else's advice, or other shortcuts, rather than on the basis of careful analysis.

The five views of the energy user draw attention to certain social and psychological processes that are significant for energy use but that have not figured prominently in past analyses of energy consumption. These include imitation; interpersonal communication from friends and others; the "momentum" of past behavior; the effects of personal choice and commitment; the expression of values; and the activation of norms. Findings from research on these processes indicate that energy policies and programs that rely for their effectiveness on the actions of a large number of people can be improved by taking into account the social and psychological processes of energy users.

We have pointed to a series of issues that any effective energy information programs must address. Essentially, an energy information program is a communication process, rather than a simple form of information transfer. This fact has implications for the way information is presented and for the design of energy programs.

To be effective, energy information programs should be based on established knowledge about communication processes. While this point seems obvious, most programs to inform individuals and households about energy saving have been poorly constructed as communications vehicles. Energy information should be presented in an attractive format to get attention, and care should be taken to make the information understandable to the intended audience. Vivid, personalized examples are useful to make information memorable, and information is presented most effectively through personal contact. In addition, information should be organized so that it is relevant to a user's specific concerns, is presented at a time and place that makes it relevant to those concerns, and is delivered by a source that is likely to attract attention and to garner trust. Because of the diversity of energy users and their needs, these requirements usually imply different messages and different information sources for different segments of the public. Information programs are likely to be more effective if they are decentralized to allow for a variety of methods of reaching diverse groups of people. It is also valuable to employ outreach workers or to rely on informal social networks within communities to deliver information. When outreach workers are used, they should receive training in communication skills and in effective methods of communicating.

Energy information is more effective when it makes energy and energy savings more visible and understandable to energy users. As was discussed in Chapter 3, to the extent energy flows remain invisible, energy users fail to respond to the signals given by rising energy prices. One way to make energy more visible is to offer frequent and meaningful feedback to energy users about the rate at which they are using energy. The typical utility bill is quite inadequate to this task, as demonstrated by the evidence of how much change can occur when feedback is improved. Cost-effective alternatives can be developed. Research is needed on how to design utility bills to be more useful as feedback, to develop ways for householders to learn more from reading their own meters, and to design feedback monitors and displays for use in residences, automobiles, and large buildings. Such research should emphasize communicating information in meaningful units and in exciting, eye-catching display formats.

The energy efficiency of equipment can also be made more visible by developing simple, understandable indices comparable to miles-per-gallon for automobiles. Further field testing of indices such as energy efficiency ratings for appliances and buildings is warranted because effective indices would decrease the level of effort needed to respond to energy price signals.

Energy audits and other informational programs can make energy flows and savings more visible by using vivid demonstrations with smoke sticks or infrared scanners and by using feedback techniques and energy efficiency indices. The question of whether the effort to make energy more visible must be publicly sponsored is still open. We do, however, have some doubts about the likelihood that utilities, which command extensive research resources and are in a particularly good position to make energy visible, will often do so effectively. They have done relatively little in this area so far, and their information programs would conflict with a desire to minimize customer complaints when feedback calls attention to rate increases.

Energy information programs must earn public trust.   This need must be addressed in the design of any program. Information can be channeled to potential audiences through sources they find credible. To maintain their credibility these sources should monitor the quality of the information. Procedures and incentives can also be created to encourage information-providers to work in the energy user's interest. It is also possible to give energy users better ways to make independent judgments—by making energy and energy savings more visible so the effectiveness of programs is more evident; by creating new institutions, on the model of *Consumer Reports* magazine, to work in the energy users' interest to evaluate information and energy-related services available at the local level; and by directing the best available information to informal social networks that are highly trusted. Some of our suggestions may also be applicable to other programs. For example, tax-credit programs and low-income weatherization services are like information programs in that they have the same trouble getting the attention of all eligible energy users.

Information by itself is not sufficient to bring about all or nearly all the energy-efficient investments that would save households money over the long term.   Some of the other significant barriers were outlined in Chapter 3, and others are discussed in Chapter 5.

Notes

1. A more detailed discussion of these issues has been presented by Schipper (1976).

2. There remains some confusion in the literature about the definitions of these influences and the relation of each to energy use. For example, Neels (1982) has concluded that in housing, "feasible changes in occupant behavior would reduce energy use only 2 to 5 percent." This conclusion is based on a regression analysis in which "behavior" was operationalized as energy price, income, household size, number of nonworkers, home ownership, and the presence or absence of an elderly household head. Neels considered these variables behavioral in that they operate through behavior, but they are not in fact behaviors. It is reasonable to expect that measured behavior would have a stronger effect on energy use than these more indirectly related influences. There is evidence to support this view (e.g., Verhallen and van Raaij, 1981). Recent research suggests that several levels of causal influence on energy use exist. Causally earlier influences may act either directly on energy use or indirectly through intervening variables such as housing stocks, general and specific attitudes, and energy-using behavior (Olsen, 1981; Stern, Black, and Elworth, 1982b; Verhallen and van Raaij, 1981).

3. A study by the Office of Policy, Planning, and Analysis (1982) of the Department of Energy found that in residential buildings between 1973 and 1980 curtailments such as temperature setbacks, room shutoffs, and reduced appliance use accounted for three times as much in energy savings as the combination of improved insulation and more efficient heating and cooling equipment and appliances. Yet most estimates suggest that the overall potential savings are greater from efficiency than from curtailment (e.g., Kempton, Harris, Keith, and Weihl, 1982; Stern and Gardner, 1981). The evidence that households' first responses have been curtailments only underlines the importance of efficiency improvements for achieving further savings.

4. The estimates of energy savings in this study were based on calculations that project the effect of energy-efficiency investments on energy use, rather than on measured energy consumption. The research can also be questioned because it was not a controlled experiment: participation in the program was voluntary, and it is possible that participants had already decided to make major investments in energy efficiency before calling the program. While this may be so, when the statistical analysis

held several economic, demographic, and attitudinal variables constant by regression procedures, the effect of program participation remained strong.

5. Survey research has usually failed to find simple correlations between general environmental concern and energy-using behavior (Farhar et al., 1979; Olsen, 1981). The new evidence points rather to indirect causality, in which general attitudes affect more specific attitudes, beliefs, and norms, with those attitudes then influencing behavior.

6. In some ways, the nuisance and uncertainty involved in shopping for home insulation is not unusual and affects many major purchases for the home. However, the existence of many untested firms in the home energy efficiency business makes anxiety realistic, and the national importance of changing energy prices makes consumer paralysis important as a public issue.

7. This can be seen as an example of a "satisficing" decision strategy, as described by Simon (1957).

8. One might ask whether it is possible for an appeal to be too vivid. The answer is almost always in the negative. The exception occurs when extreme fear is being created, such as when gory pictures of cancerous lungs are used to discourage cigarette smoking. Extreme fear can, under certain specifiable conditions (Higbee, 1969; Leventhal, 1970), tend to immobilize an individual. This is an extremely unlikely occurrence, however, with appeals to save energy.

9. There is a large research literature in social psychology bearing on dissonance phenomena: Irle and Montmann (1978) list 856 separate published articles (largely research publications).

10. Some studies show that energy use was responsive to general social conditions, apart from price, during the 1973–1974 oil embargo (Walker, 1980). Exhortations from government and a general crisis atmosphere probably combined to influence the public at that time.

11. This criterion assumes the view of energy as commodity. Other views of energy imply other goals for government programs— the view of energy as ecological resource, for example, seems to imply encouraging energy users to minimize energy use beyond that dictated by their individual economic self-interest.

12. The data are not always consistent, however; Hutton's (1982) review of three recent conservation campaigns found no evidence for a linear relationship between repeated exposures of a communication and increased effectiveness.

13. A payback estimate is obtained by dividing the cost of an investment (e.g., a storm window) by the value of the energy it saves in a year. Although it is an imperfect guide to economic self-interest, it is potentially usable by an average householder. However, it is only one of several possible simple indicators (e.g., annual savings, percentage return on investment), and there is no research comparing the usefulness of different indices.

14. Some variables that might also thought to be important have not been shown to relate to the effectiveness of feedback; these include the use of mechanical versus human sources of the information and, within some range, the units in which the information is presented. Feedback has been about equally effective when delivered in a variety of units: temperature-corrected consumption units, energy use compared to similar households, temperature-corrected dollars-per-day, and so forth. The credibility of the information seems more important than the particular units (Seligman et al., 1981).

15. Problems can arise when the initiative is left to the utilities. In 1974, a New Jersey electric utility started informing residential customers of their present electricity consumption in comparison with weather-corrected consumption from the previous year. The program was initiated to promote energy savings, but it was soon cancelled. What happened was that after a rate increase, many customers used the new information to bolster complaints that even though they were using less energy than before, their bills had increased. Thus, partly *because* the program succeeded in reducing energy use (Russo, 1977), it was discontinued. Customer reaction might have been different had the utility added information on what the bill would have been without the conservation efforts. As it was, the bill made energy savings visible, but not in the units that were most meaningful to the utility's customers.

16. While the RCS is more than just an energy audit program—in fact, the energy audit may not be the most important features of the RCS package (Stern et al., 1981)—and RCS energy audits are administered in many different ways around the country, the home energy audit as described in the RCS regulations is prototypical of energy audits in the United States.

17. While the RCS regulations were later revised (*Federal Register*, June 25, 1982), the 1979 requirements were still being followed in most programs in late 1982 (according to M. Friedrichs, of the Office of Building Energy Research and Development, U.S. Department of Energy).

18. In most simple models of rational economic behavior, the cost of travel is the same in both instances, as is the benefit ($5); thus, the cost-benefit ratios are the same. That one item is marked down 33 percent and the other only 4 percent does not make one trip more worthwhile than the other when economic savings is the sole criterion of value.

19. For more detailed discussion of some behavioral issues in these domains, see Stern, Black, and Elworth (1981).

5

Organizations and Energy Consumption

Organizations affect energy consumption in at least four major ways. They transform one form of available energy into another: for example, an electrical utility company turns coal into electricity. They consume energy in their own activities, including their production processes: for example, a metal processing firm. They produce products that consume energy: for example, an automobile manufacturer. And they make choices that determine energy use for others: for example, an architectural firm. Many organizations, of course, do all four; thus, organizational behavior affects the aggregate efficiency of energy use. This chapter examines this relationship, drawing on research studies of organizations. The first part of the chapter focuses on organizations as energy users; the second on the roles of intermediaries—many of which are organizations—in energy use.

ORGANIZATIONS AS ENERGY USERS

There is very little empirical research that focuses directly on organizational behavior with respect to energy consumption. Although some organization-level data have been collected by the Department of Energy and other federal agencies, the quality of the data for research purposes is not established, and the data have been little used to explore systematic variations among organizations. There are also many reports of individual organizational efforts to improve energy efficiency; those reports are useful to provide hints about what is going on, but they are both too fragmentary and too impressionistic to yield a clear picture. As a result, what we say is speculative, based more on general knowledge about organizations and on inferences from the rather casual, available accounts of energy programs in organizations than on firmly established empirical observations.

A conventional way to think about energy use in organizations is to treat organizations as simple, self-interested rational actors. The most common assumption about the private sector is that firms attempt to maximize profits. In the public sector, the most common assumption is that agencies try to maximize size or political support. In this view, energy decisions are like any other organizational decisions: energy conservation will occur when the expected costs of conservation are less than the expected gains to be realized. It is assumed that firms will invest in energy efficiency when it is profitable and that agencies will invest in energy efficiency when it contributes more than it costs for organizational growth.

The hypothesis that organizations are rational actors is useful for predicting and interpreting organizational behavior in the aggregate. According to the hypothesis, an increase in the cost of energy for industrial firms, without a comparable increase in the cost of more energy-efficient equipment, will increase the total industrial investment in such equipment. Similarly, an increase in the cost of energy for public agencies, without comparable increases in total budgets, will increase the agencies' investment in energy efficiency. Aggregate statistics, as well as numerous case reports, do document that organizations, as a group, are responsive to substantial shifts in energy costs. In particular, major increases in the price of energy and incentives in the tax system have produced organizational response. For example, Hsu (1979) found that the expected payback period was an important factor influencing the adoption of energy-conserving technologies. Behavioral research suggests, however, that as with individuals, aggregate statistics about industry or government conceal considerable variation.

Energy use among firms and among agencies varies in ways that may be due to systematic features of organizational structure and behavior. In particular, studies of organizational decision making identify two major features of organizations that affect the fit of a simple rational view to their actions. First, an organization is not a single actor with a single objective, but a collection of actors with potentially conflicting objectives. Although the existence of an organization suggests some mutual gains from cooperation, not all issues of conflict are resolved by explicit or implicit contracts among the actors. The interests of one department or employee may conflict with the interests of others, and those conflicts affect the way in which an organization makes decisions. For example, workers with innovative ideas for changes in technology often require the acquiescence, or even collaboration, of managers and others in order to implement the change.

Second, organizations, like individuals, do not act on the basis of complete and precise information, but respond in a confusing world of vague, amorphous stimuli, unclear consequences, and ambiguous goals. Rational action is based on two assumptions about the future: one about the probable future consequences of current action and one about the future preferences in terms of which of those consequences can be evaluated. These are difficult assumptions to make about any organization. For example, the consequences of action may depend on the not easily predicted actions of others, and future preferences may not only be different from current preferences but may also be affected by actions taken.

These features of organizations lead to regularities in organizational actions that are not immediately obvious from the rational model and that may influence decisions affecting energy efficiency. They can be seen as amendments to, rather than replacements of, the simple rational model. Consideration of these factors may help policy makers understand some of the frictions in organizational response to conventional financial incentives and regulations, as well as the circumstances under which a given organization might invest less (or more) in energy efficiency than it would be expected to from the point of view of organizational self-interest.

Success, Failure, and Organizational Slack

When an organization is doing well, in its own terms, it acts differently than an organization that is doing poorly. If targets (for example, profit goals) are being met, policies and explicit strategies become more risk-averse, and the search for refinements in current activities becomes less important; but because internal control is relaxed and organizational slack increases, practice becomes more risk-seeking. On the other hand, if targets are not being met, policies and explicit strategies become less risk-averse, and the search for refinements in current activities becomes more important; but because internal control is tightened and organizational slack decreases, practice becomes more risk-averse (Cyert and March, 1963; March, 1981; March and Shapira, 1982).

These theories about risk taking and success have been developed as a result of apparently contradictory arguments and data on the effects of performance on innovation. There are data supporting the idea that innovation is stimulated by adversity,[1] while other studies either do not support that hypothesis or suggest that innovation is more characteristic of successful organizations.[2] Such contradictory results have been interpreted in terms of a distinction between "problem" search and "slack" search (Cyert and March, 1963). The former is characteristic of unsuccessful organizations and leads to refinements of existing technologies; the latter is characteristic of successful organizations and leads—under some circumstances—to new technologies. Although there are variations in discussions of the relation between organizational slack and innovation, it is generally argued that success and the presence of slack lead to a loosening of controls and decentralization (Pfeffer and Leblebici, 1973; Pfeffer, 1978) and that failure (Perrow, 1981) or stress (Staw, Sandelands, and Dutton,

1981; Hall and Mansfield, 1971) leads to a tightening of controls and centralization. Under conditions of organizational slack, lower-level decision makers are likely to take actions that are "foolish" from the point of view of general organizational objectives but have a chance of being serendipitously innovative (March, 1981). As a result, decentralization leads to greater innovation (Wilson, 1966; Zaltman et al., 1973; Moch and Morse, 1977; Pierce and Delbecq, 1977; Daft and Becker, 1978), and slack search contributes to the discovery of new technologies. For example, Hsu (1979) found that the adoption of energy-saving technologies was related to organizational slack generated by rapid growth.

Such a theory is different from a simple rational model. In a rational model, neither organizational performance targets nor the current level of performance—as seen in the current level of profits—is relevant to understanding or predicting decision behavior. That is, the investment decisions of an organization should be unaffected by its current subjective success. Studies of organizational behavior, on the other hand, seem to indicate that organizations that are achieving their goals will be less likely to search for energy-saving modifications of current technologies and procedures and more likely to search for state-of-the-art modifications that cater to the professional training and attitudes of subunits or individuals without necessarily making immediate contributions to manifest organizational goals. Unsuccessful organizations are likely to have shorter time horizons than successful organizations; thus they will be more interested in retrofits or refinements than in major changes and less likely to invest in technologies with long payback periods. Conversely, successful organizations are more likely than unsuccessful firms to take actions that have low probability of success but are characterized by some chance of major gains.

Rules, Routines, and Budgets

Organizations follow rules. Every organization has a large collection of rules that direct and coordinate activities, and most decisions are the consequence of applying a set of rules to a situation. Rules change over time, are often violated, and are frequently inconsistent. But most actions in an organization are not the result of decisions in which alternatives are considered in terms of their consequences for prior objectives—most actions involve finding an appropriate rule and using it. Rules in organizations specify routines. Without routines, it would be hard to accomplish many of the things that are done easily within modern organizations.[3] Organizations generally solve problems and respond to environmental demands by applying existing routines rather than by developing new ones (Allison, 1971; Halperin, 1974; Hall, 1976; Perrow, 1981).

Rules are broadly adaptive, but there is no guarantee that they reflect optimal, or even acceptable, solutions when new problems arise (Altman, 1971; Watney, 1974). Some activities reflect professional standards or traditions rather than an organization's specific work situation (Rohr, 1981). Sometimes a routine that has developed historically continues to be used even when more useful routines are known to exist (Baran et al., 1980). Sometimes a rule is insensitive to situational factors (Argote, 1981). Sometimes existing rules make risky or novel projects difficult (Chakrabarti and Rubenstein, 1976). Furthermore, appropriate rules are not always followed. Individuals do not always know their organization's routines accurately (Sproull, 1981b), nor do they always observe them (Ellsberg, 1972; Britan, 1979; O'Reilly and Weitz, 1980; Sproull, 1981b). Managers sometimes do not correctly perceive the environment, and thus select the wrong rule (Wohlstetter, 1962; Starbuck et al., 1978; Perrow, 1981), and they sometimes fail to set the goals required for rules to function (Mowery et al., 1980; Lowenthal, 1981).

Although rules and routines are stable in the short run, they change.[4] When rules change, the change is likely to be local to the problem area (Cyert and March, 1963; Perrow, 1981) and not rationalized with the rest of the organization's procedures (Hall, 1976).

The most obvious organizational rules are those associated with budgets, which represent plans and agreements. Organizations use budgets and rules about them to manage expenditures. They can be made somewhat flexible and somewhat contingent on uncertain future events, such as revenues, and they may be renegotiated to some degree, but budgets function as routines for delegating expenditure authority. They are rarely underspent, relatively rarely overspent by much.

Expenditures that can be fit into a current budget require less organizational consultation and approval than those that cannot. As a result, the real availability of funds for a project in an organization depends on such things as the stage of the budget cycle, the departmental location of the project, and the amount of slack in the budget. Many organizational investments relevant to energy consumption involve asking whether there is money in the budget for the project, rather than what its return on investment or payback period might be.

Other relevant rules for understanding organizational behavior with respect to energy efficiency are accounting rules and rules of architectural and engineering design. Decisions on capital investment, including investment in energy-efficient equipment, depend on an assessment of the relative value of alternative investment opportunities, and that assessment follows standard accounting procedures for determining and allocating costs and returns. For example, the relations among initial capital costs, costs of maintenance, and costs of operation in the accounting of project costs can be important to a decision. So-called life-cycle accounting, in which operation and maintenance costs play a significant role, ordinarily increases the likelihood that energy-efficiency considerations will be important. Similarly, engineering standards, design specifications, and building codes that emphasize BTU/cost ratios or other energy considerations increase attention to such factors. Rules such as these, which emphasize energy considerations, have the effect of making energy more visible to critical personnel in the organization, as discussed in Chapter 3.

Avoidance of Uncertainty

Organizations value certainty; they try to avoid uncertainty. When consequences are uncertain, managers are likely to delay action. Thus, Webster (1971) found that the expected profitability of an innovation is a less significant factor in a firm's adoption decision than the degree of perceived risk. It is possible to interpret the behavior as risk aversion, but the main point is not that organizations avoid risk in choosing among alternatives; rather, in their effort to reduce error in projecting possible outcomes, they will even change the nature of the alternatives before them. Managers do this partly by emphasizing short-run responses to short-run feedback rather than trying to make long-run forecasts or to base action on them. They also seek to gain control over critical resources in the environment through contracts that ensure reliable supplies and through vertical integration. They try either to control the suppliers of vital resources or to buffer their dependence on them through inventories. It appears to be true that organizations, particularly successful organizations, will characteristically be willing to trade some expected value in order to sustain control and avoid the risk of an unfavorable outcome.

Organizations make efforts to control the uncertainty that exists because of their dependence on other organizations (Emerson, 1962; Blau, 1964; Thompson, 1967; Jacobs, 1974; Pfeffer and Salancik, 1978). For example, there is evidence that purchasing decisions in organizations are made in a way that minimizes uncertainty, subject to a budget constraint, rather than in a way that involves explicit trade-offs between risk and cost (Dickson, 1966; Lehmann and O'Shaughnessy, 1974; Kiser and Rao, 1977; Corey, 1978). Their efforts to avoid uncertainty in their environments include such broad strategies as vertical and horizontal integration, diversification, interlocking directorates, joint ventures, and interorganizational coordination.[5] In a similar fashion, organizations seek to avoid uncertainty internally through incremental decision rules (Lindblom, 1959; Padgett, 1980a, 1980b), by following rules (Cyert and March, 1963), by building inventories of inputs and outputs (Thompson, 1967), by loose coupling (March and Olsen, 1976; Weick, 1976), and by organizational slack (Cyert and March, 1963; Lowe and Shaw, 1968; Schiff, 1970; Schiff and Lewin, 1968; March, 1981).

One obvious implication for energy decision making is that organizations will be attentive to issues of energy efficiency only after they have protected themselves as much as possible from energy dependence. For example, a study of major retail chains showed that chains were more likely to take energy management steps when they believed their energy futures were uncertain and when they lacked trust that government programs would help them (Mills, 1981). Organizations also will be more sensitive to the possibility of a supply cutoff than to price changes. A study of energy conservation among thirty-two firms found that firms in the New York City area, which had experienced acute shortages of natural gas, were significantly more likely to have made investments in energy conserving technologies than were similar firms in the Allentown, Pennsylvania area (Hsu, 1979).

Organizations are also sensitive to energy-related disruptions of business operations. Among retail chains, for example, firms that had experienced shortened hours, employee and customer inconveniences, and other disruptions as a result of energy conditions did more to manage energy in their operations than firms that had not experienced such disruptions (Mills, 1981). Organizations probably will be responsive to energy options that make their own supplies of energy relatively independent of others—for example, the cogeneration of heat and electricity—even if the options are not particularly good in terms of a cost-benefit analysis. As energy costs increase, managers will look for possible mergers or other forms of organizational arrangements that give them greater immediate control over their own supplies. If energy shortages seem likely, they will try to build their inventories and will probably build them to levels that are not justified by standard decision analysis.

Scarcity of Attention

Attention is as scarce a resource in organizations (Simon, 1971, 1973) as it is for individuals. An organization can be viewed as a collection of problems looking for solutions and a collection of solutions looking for problems to which they might be applied. Problems and solutions seek attention from decision makers. It is a commonplace observation that action is more likely in an organization if it can command the attention of high-level managers (Chakrabarti and Rubenstein, 1976; Hsu, 1979). But since decision makers have limited time and capacity for attention, not all alternatives are considered; not all information is gathered; not all available information is considered; not all values are made conscious. So, the process by which attention is allocated is a key part of any decision.

One way attention is allocated is by routines. For example, in developing budget decisions, routines lead organizations to focus on budget base and increment (Kamlet and Mowery, 1980), on estimated revenue (Larkey and

Smith, 1981), or on budget targets (Padgett, 1981). "Garbage can" decision routines (Cohen, March, and Olsen, 1971) connect problems and solutions on the basis of their simultaneity, rather than their substance. Routines for the gathering of data tend to accumulate so much information that important pieces are not noticed (Wohlstetter, 1962; Turner, 1976; Perrow, 1981). Information of importance is likely to be overlooked if it is not attended to on the basis of standard organizational rules (Ellsberg, 1972; Neustadt and Fineberg, 1978) or standard rules of professional practice (Dearborn and Simon, 1958; Perrow, 1981).

Attention is also influenced by the ways in which managers allocate their time and attend to information. Managers tend to communicate orally rather than in writing (Mintzberg, 1973; Keegan, 1974; Lyles and Mitroff, 1980; Sproull, 1981a) and appear to respond to attention requests more in terms of the source (Cohen and March, 1974) or the medium (Sproull, 1981a) of the request than its content. As a result, the network of oral contacts among acquaintances is an important form of environmental scanning (Keegan, 1974; Tushman and Scanlon, 1981). Managerial attention is brief, frequently interrupted, and as likely to be initiated by others as by the manager (Dubin and Spray, 1964; Mintzberg, 1973; Cohen and March, 1974; Sproull, 1981a). Managers are most certain to attend to disruptions and emergencies.

Managers attend to information sources selectively, recording information from trusted sources (Pettigrew, 1972; Neustadt and Fineberg, 1978) and ignoring information from unfamiliar or untrusted sources (Wohlstetter, 1962; Turner, 1976). They focus on familiar aspects of complex stimuli, ignoring less familiar features (Dearborn and Simon, 1958; Turner, 1976). Managers use relatively simple models for identifying problems (Pounds, 1969; Sproull, 1981b) and for evaluating decision alternatives (Cyert et al., 1958); they are influenced by broad organizational expectations and beliefs (Mintzberg, 1978; Starbuck et al., 1978) and by personal goals (Pettigrew, 1972).

In any organization there are energy problems looking for solutions, and there are energy solutions, including conservation, looking for problems, but energy must compete for attention with other problems facing the organization and other solutions being offered. Thus, the amount of attention devoted to energy efficiency depends on the number and salience of other, competing issues that demand attention and time. The salience of energy (and other issues) depends on several organizational factors. For example, it depends on conspicuousness, indicated by the amount of money devoted to it or the number of functions it affects. This, in turn, depends in part on the ways in which organizational accounts are organized and the kinds of scorecards that are kept on organizational performance. Salience also depends on routines: standard operating procedures dictate much of the allocation of attention. Finally, attention depends on the structure of interests and power in the organization. Organizations attend to things that are on the agendas of active, well-organized interests, and those interests can be either external to an organization or a part of it. If there is a subunit whose survival and growth depends on bringing energy solutions, that person or group will help to discover an energy problem in the organization and make it salient to others.

Diffusion and Imitation

Organizations imitate other organizations. A large fraction of the innovations introduced in any organization are copies of what is being done elsewhere. This includes most changes in standard rules of "good practice"—for example, good engineering practice, good accounting practice, or good managerial practice—for which trade and professional associations often provide a not entirely unbiased mechanism for spreading beliefs and procedures from one organization to another. It also includes technical and managerial innovations, which decision makers discover and justify by noting what is being done by others.

The empirical study of imitation in organizations is complicated by the difficulty of distinguishing imitation from independent, but subsequent, adoption of a practice. Studies of the diffusion of new technologies often show a pattern of adoption that is consistent either with a contagion model or with a normal distribution of independent adoption times.[6] There is some more direct empirical evidence for the imitation hypothesis: for example, personal communication among peers in similar positions in different firms speeds the adoption of a practice (Trippi and Wilson, 1974; Czepiel, 1974; but see Webster, 1970). In the flour milling industry, innovations spread more rapidly to mills that are part of groups than to independent mills (Hayward, 1972). Studies of information seeking about innovation indicate that information is often sought through interpersonal channels (Trippi and Wilson, 1974), particularly from vendors and suppliers (Webster, 1970; Ozanne and Churchill, 1978; Mills, 1981). There also appear to be industry opinion leaders who influence adoption decisions (Carter and Williams, 1959; Webster, 1970; Czepiel, 1974).

Thus, it appears that imitation among organizations follows a pattern not totally unlike imitation among individuals. It is sensitive to factors of contact (which organizations are connected with which others), transmission (what kinds of information about organizational practices are communicated), source (the credibility of the information provider), and receptivity (what makes something being done in one organization attractive to another). Contagion is particularly likely when it is difficult for an organization to assess the appropriateness of an innovation or when there are ambiguities about objectives. Under those conditions, action is made to seem sensible to one organization by its adoption by others: one would expect the relative standing of two organizations to affect the rate at which one imitated the other, and one would anticipate the development of fads in organizational decisions.

Policy Principles to Affect Organizational Energy Users

It is easy enough to see that any of these features of organizational decision making may lead an organization to invest less in energy efficiency than it should, even from its self-interested point of view. Under such circumstances, public policies might be directed toward overcoming these limits on rationality so that organizations act to improve their own positions, as well as to be in conformity with national policies. In such a situation, everyone gains and no one loses. If a political decision is made to consider energy efficiency a national interest, it would also be possible to induce organizations to undertake energy decisions that are not in their narrowly defined self-interest. In either case, the point is to lead organizations to do things they would not do without interventions.

Such problems have traditionally led to various kinds of "carrots and sticks" designed to change behavior. These include taxes, subsidies, regulations, and threats. The major advantage of using taxes and subsidies hinges on their effectiveness in manipulating organizational actors into compliance with public policy objectives in an efficient way. When organizational actions are affected by any of the factors that have been discussed, the timing and form of governmental incentives and regulations may be important considerations in influencing organizational actions, and taxes and subsidies of the usual sort may be much less effective than hoped. In addition, policies within organizations or in professional and trade associations may also be effective.

The general principle underlying effective policies is that, to ensure that incentives for energy efficiency influence organizational action, energy efficiency must be interpreted into organizational goals, procedures, and routines. This can be done by assigning responsibility, building energy considerations into accounting and design routines, improving measurements of energy flows, and by using the resources of trade and professional associations. In order to improve energy efficiency in organizations, it is particularly important that high-level managers identify energy efficiency as an important organizational goal; this attracts attention to energy within the organization. To improve energy efficiency, at least one person at a high level in the organization should be made responsible for advocating energy efficiency and measuring progress toward concrete objectives. To the extent that energy objectives can be specified for particular activities, tasks, or subunits within an organization, attention to energy problems will filter down through the organization.

Procedures such as separate budgeting of energy costs and life-cycle accounting for energy-using capital investments build energy concern into the organization by routinizing it. Other rules and routines in areas such as architectural and engineering design can have the same effect. Energy efficiency can be used as a criterion of effectiveness in measuring and allocating organizational resources. For example, in planning budgets, a separate category of capital outlays for energy efficiency can be created. Or, in evaluating the productivity of labor, the advantage of energy savings can be assessed.

To identify potentials for energy efficiency and to measure progress, organizations need devices and indices for monitoring energy expenditures. They should keep simple, cumulative records of the data generated, make these records readily retrievable, and statistically evaluate the payoffs of past investments in energy information and energy-saving measures and technologies. With small computers, even a fairly small organization can evaluate its expenditures and investments, taking into consideration external factors such as weather, prices, and competitors' behavior.

Trade and professional associations are valuable sources of information on effective ways of saving energy. These groups are in a good position to tailor information to their members' needs; they are also highly credible and can quickly spread new ideas to places where they are most likely to be applicable.

Trade associations and associations of public officials can assist their members in improving energy efficiency by establishing regular networks within the associations within which organizational representatives concerned with energy can interact. They can also gather, digest, and distribute brief reports about successful energy-efficiency efforts of well-respected firms or agencies. Such procedures would assist the process of social diffusion.

The U.S. government is the largest organizational energy user in the country. As such, it can implement many of the above suggestions. It can, for example, assign more resources to managing its own energy use, delegate responsibility for energy efficiency to high-level officials in each department, and set specific goals for energy saving in each agency.

The federal government should work to develop ways organizations can improve their energy efficiency. A research effort would begin as part of a drive to improve energy efficiency in government operations. It might initially focus on the effects on organizational energy use of engineering decisions, accounting and monitoring systems, and the organizational status accorded energy management. Policies found to be effective in federal agencies are often adaptable in state and local government agencies and in organizations in the private sector. The research effort might be broadened to include cooperative work with private sector organizations so that further transfer of learning will occur. The federal government can take the lead by spreading the knowledge it develops, distributing reports through professional associations or through the mass media. More aggressive federal policies for organizational energy efficiency are conceivable. For example, municipalities have been required to designate city energy managers, a requirement consistent with our suggestions. Given a higher national priority on energy efficiency, the federal government might also offer incentives or even set requirements for energy conservation by private organizations.

All these suggestions involve using knowledge about organizations to lead them to be more attentive to energy efficiency. The rough clues that we have provided suggest that government and private sector initiatives can be strengthened by some moderate attention to the ways organizations select and pursue performance targets; to the effects of organizational slack; to the pervasiveness of rules and routines (particularly budgets and other accounting rules); to the extent to which organizations seek to avoid uncertainty; to the complications of attention scarcity; and to the ways in which ideas and practices are diffused.

THE ROLE OF INTERMEDIARIES IN ENERGY USE

Energy use does not depend only on the behavior of energy users. Actions that do not themselves involve the use of energy can have profound effects on energy use. This section discusses the roles of some of the important actors that function as intermediaries in energy use by making choices for the ultimate energy consumers. Intermediaries are organizations, individuals, and members of professional reference groups who stand somewhere between the orginators of energy-related goods, services, and information and the ultimate energy users. Intermediaries warrant discussion separately from energy users because the interests of the two groups are not always the same: conditions that may motivate an individual or organization to use less energy will not necessarily motivate an intermediary to act in ways that will decrease energy use.

Purchasers of Buildings and Energy-Using Equipment

Buildings and energy-using equipment are often purchased by persons or organizations that will never use them or will never pay for the energy they do use. About one-third of the housing units and more than one-half of the commercial space in the United States are rented. Furthermore, 40 percent of all major appliances are purchased by builders (Science Applications Inc., 1982), and the percentage is probably higher for furnaces and central air conditioning systems, which are almost always built in. So, at least half the energy used in U.S. buildings is probably used in buildings or by space and water heating equipment originally purchased by intermediaries. In the transportation sector, a significant though probably much smaller proportion of cars and trucks was originally purchased by rental agencies or fleet operators. Thus, much of the nation's stock of energy-using equipment was purchased by intermediaries. Consequently, much of the market pressure on manufacturers of energy-using equipment comes not from ultimate energy users but from the intermediaries who purchase their products with no expectation of paying for the energy those products use.

While the interests of ultimate energy users are consistently related to minimizing long-term energy costs, the interests of intermediaries are more variable and complex. Ultimately, intermediaries who build structures or purchase energy-using equipment for resale or rental are interested in making their products attractive to customers, in maintaining their reputations, and in minimizing the costs associated with repairs and with underutilized equipment. Energy efficiency becomes a priority to the extent that the buyers or renters of their buildings and equipment consider it in their decisions. But because energy is only one of many aspects of any product and because energy efficiency, especially for buildings, is often hard to determine in advance, purchasers or renters who want energy efficiency may not show it in their choice of products.

Thus, a car rental agency may buy more large-size cars than it needs because it is less likely to lose business by giving a customer a larger-than-requested car than by offering a smaller-than-requested model. For the customer, availability of the car is more important at the time of rental than energy use. For these reasons, gasoline consumption of rental cars is higher than would be desired by the individuals or organizations that pay for the fuel. A home-building firm may spend extra money on brand-name kitchen appliances because prospective buyers will take this as an index of the quality of the home (Quelch and Thirkell, 1978). When purchasing a furnace—which most home purchasers do not examine—the same home builder will be more concerned with low price and quick delivery than with quality. Energy efficiency will not be a major criterion in either choice unless home buyers begin to demand it. For such demand to develop, home buyers must be well informed about energy efficiency in appliances. But they may be unlikely to gather such information at a time when they must also pay attention to issues of overall cost, design, and durability of their prospective homes. As a result, energy efficiency will often tend to be undervalued, and it almost certainly will not be accurately judged.

Moreover, energy-efficient innovations are risky for intermediaries. For example, in the construction industry, builders assume large risks by adopting innovations in solar design or other energy-efficient practices. They must change their procedures, by hiring design professionals or new subcontractors; they may need to make special efforts to satisfy building codes; and ultimately, buyers may not value the innovations enough to repay the financial investment and the difficulties involved in innovating. An intermediary's risks are different from, and may sometimes be greater than, the risks of an energy user who adopts a new energy-saving technology. It may take strong evidence of buyer interest to get a builder to take the risk of changing past practices. This is a particular problem because builders' perceptions of their markets may lag behind buyer preferences. One study of home builders and potential homeowners showed that builders were much more pessimistic about solar heating than were the prospective buyers who were shopping for a home (OR/MS Dialog, 1980).

The Problem of Rented Buildings. Probably the most significant example of the gap between purchaser and user is in rented space. Here, the building owner is a critical intermediary. The owner's interests emphasize the durability and reliability of equipment as well as initial price (Quelch and Thirkell, 1978; Office of Technology Assessment, 1980); energy is not a major concern. And the path from the energy user's interest to the purchase of equipment is more tortuous than in other situations, because there are two intermediaries—the builder and the building owner—each with interests different from the ultimate energy user's.

Renters, whether they are households or organizations, are limited in their options to use or save energy. They can make behavioral changes, such as shutting off air conditioners and lights, and they may make small changes in the building, such as weatherstripping around windows. But they are unlikely to make larger investments (Stern, Black, and Elworth, 1982a), which may be construed as tampering with private property; even if allowed, these investments are unattractive since they improve someone else's property. In addition, some renters are limited even in their behavioral options—for example, by central control of heat in an apartment or office building.

For occupants of rented buildings, the problem of invisibility (see Chapter 3) is doubly acute. Renters rarely purchase the heating and cooling equipment they use, are not responsible for its upkeep and operation, and in many cases never even see the bills for energy use. As a result, it is even harder for a renter to understand the energy flows in a building than it is for an owner-occupant. Because of this, even if renters were given incentives to cut energy use, they would probably have few ideas about how to earn them.

Owners of rented buildings are also constrained in various ways from improving energy efficiency. In a new building, costs of construction (and therefore rental costs) can often be kept down by installing electric resistance heat, which is not the most energy-efficient plant for an entire building. Thus, competition for tenants may lead to energy inefficiency, especially when tenants pay for heat but are not particularly concerned about heating costs at the time they choose the space. After a building is occupied, the owner still has limited incentive to improve its energy efficiency. If the occupants pay for heating and cooling, they get the immediate benefits of any investment, even though the owner will eventually benefit if prospective future occupants consider energy costs. If owners pay for heating and cooling, they may be deterred from investing by the prospect that the occupants will use up the potential savings by wasteful behavior, even though this would not happen unless the occupants became more wasteful after the investment than they were before. In general, then, owners and occupants can prevent each other from benefitting from certain energy-saving actions.

Policy Alternatives.   Some policies have been tried for promoting energy efficiency in rented buildings, and they offer hope in certain situations. For example, in master-metered buildings, which account for over half the energy used in multifamily housing (McClelland, 1980), the various methods that have been tried for providing incentives for residents to use less energy include contests among residents, rebates for energy savings, experiments with individual metering and submetering, and a residential utility billing system (RUBS) that allocates energy costs among residents by a mathematical formula. Individual metering and RUBS are the systems with the greatest potential for widespread application. Both shift the immediate economic incentives for energy saving from owners to occupants, and energy savings have been reported. An analysis of three experiments with individual metering, for example, concluded that it led to savings of about 22 percent of electricity and used for cooling, lighting, and appliances, about 14 percent of electricity in all-electric buildings, and about 5 percent of energy used only for heating and hot water (McClelland, 1980). RUBS, which does not involve capital costs for meters, produced smaller savings—8 percent, 5 percent, and 5 percent, respectively, for the three patterns of use. Both systems produced only modest savings in energy used for space and water heating which accounts for more than three-quarters of energy use in multifamily housing (McClelland, 1980).

Individual metering removes some of the incentive for building owners to invest in energy efficiency, so it may also waste energy while it saves. There is little empirical evidence available, however, on the extent to which the energy efficiency of individually metered buildings may decline. One econometric study (Neels, 1981) found that owners of individually metered buildings spent about $30 (15 percent) less per year on repairs than owners of master-metered buildings. Because more than half of this difference was made up by increased spending by tenants, the author of the study concluded that the net incentive is relatively minor. Clearly, more research is needed before drawing conclusions about the net energy savings that can be expected from individual metering or about the net distribution of costs between owners and residents that is likely to result.

Other programs and policies have also received attention. For example, utilities and commercial groups have developed certification programs (McClelland, 1982) to help building owners advertise energy-efficient buildings to potential renters. In a slack rental market, this may reward the owners of energy-efficient buildings with higher rents or lower vacancy rates. In a suburban area of Atlanta, Georgia, where much new housing is being constructed, virtually all new buildings are receiving certification by Georgia Power, and most are advertising energy efficiency to attract renters.[7] Local ordinances have also been proposed to require that buildings be brought up to an energy standard at resale. These ordinances make sense in that enough money for upgrading a building is usually available at the time of resale, but the ordinances are vigorously opposed by real estate interests as an impediment to transactions.

Each of these proposals has some possibilities, but also problems. In the tight rental housing markets of many cities, it may be possible to pass energy costs along to renters with little fear of lower occupancy rates. Where there are rent control ordinances, a pass-through provision is often needed to prevent bankruptcy among building owners and the abandonment of housing stock. Owners are also frequently constrained by lack of capital, and find energy investments expensive when short-term debt financing costs are included (Office of Technology Assessment, 1982). Most owners of rental housing are small businesses with limited access to cash or credit. In addition, the poor condition of the rental housing industry makes it a poor risk for lenders. And many owners do now expect to recoup energy investments when selling a building (Bleviss, 1980).

It is difficult to generalize because of the extreme diversity in the rental housing industry. Conditions of housing markets and local ordinances vary greatly (Bleviss, 1980), and the behavior of owners varies with their situations. One study (Neels, 1981) found that buildings partly occupied by their owners used about 25 percent less energy than buildings with absentee owners in 1978, and an even greater discrepancy was projected for the future. The same study found that buildings owned by corporations and partnerships used 9 percent less energy than buildings owned by sole proprietors. The differences were attributed to increased expenses for labor and repairs among owner-occupants and to better information among "professional landlords." The importance of access to capital may also vary greatly—the limited research to date conflicts with regard to the importance of this factor as a barrier to investment.

Diversity presents a problem in another way. There is a great variety in the residential housing stock—especially among multifamily buildings—so accurate information is not available on the energy savings that can be expected from particular investments. Data are scarce, but an evaluation of data primarily from commercial buildings (Office of Technology Assessment, 1982) showed that actual energy savings varied enormously, and were often as much as 50 percent more than or 80 percent less than predicted. Until more reliable information is available for building owners, underinvestment in energy efficiency is an understandable strategy for avoiding wasted expense.

Still, because many rental housing units are highly energy inefficient, it is in the interest of both owners and occupants to devise ways to share the costs and savings. Some model contracts are being developed in commercial buildings that allow owners to pass through a portion of the capital improvement costs that will benefit occupants.[8] In multifamily housing, this procedure may prove less attractive because of mistrust between owners and occupants or because of lack of technical sophistication. However, the development of such procedures may be a ripe area for exercise of the skills of individuals and groups that have arbitrated labor disputes or mediated consumer or environmental disputes.

Operators of Energy-Using Equipment

Some large apartment buildings and many commercial buildings are operated by someone who is neither the owner nor an occupant. Truck and bus drivers and airline pilots are in a comparable situation in that they operate large energy-using machines although someone else pays for the fuel. The interests of such intermediaries may differ from those of owners and occupants or clients. Building operators or managers, for example, are motivated to minimize malfunctions of equipment or other sources of complaint. For example, a building operator may run an inefficient furnace in preference to risking a shutdown for repairs during business hours. Or a building engineer may regulate temperature to minimize complaints. This means plenty of heat in winter and plenty of cooling in summer—enough to keep down the complaints. And it will probably mean too much heat and cooling if occupants can adjust to the conditions by opening windows without notifying the engineer. This brief overview suggests special problems in achieving energy efficiency in buildings in which there are three interested parties—owners, operators, and clients. The other side of this argument is that professional building operators may use energy more efficiently because of superior technical knowledge. For operators to make use of this knowledge, however, it may be necessary to build in incentives for them to use energy efficiently.

Organizational Energy Users and "Embodied Energy"

Individuals use energy indirectly when organizations use energy to provide consumer goods and service. Organizations make energy decisions for individuals and households when they choose equipment and processes to use in manufacturing, construction, and agriculture. This energy use is "embodied" in the automobile, building, or food product when it is purchased. When markets are highly competitive and energy accounts for a substantial portion of a product's price, purchaser demand will be relatively responsive to the cost of embodied energy. But when such conditions are not met, it is more difficult for purchase decisions to influence producers' decisions about embodied energy. The energy costs of producing a soft-drink container or a bomber are significant, but because the markets allow purchasers limited choice, legislation and regulation seem to be the only effective ways to improve energy efficiency. And the costs of embodied energy in many other commodities are miniscule for the purchaser. The energy costs are often significant to the producers, but, for reasons mentioned in the discussion above of organizational energy users, organizations do not always make economically rational choices about their energy use.

A telling example of embodied energy concerns the use of energy for making or transforming energy. The "fuel adjustment" provision common in utility regulation allows utilities to automatically increase gas or electric rates when there is an increase in the cost of fuel to the utility. This arrangement gives the utility virtually no incentive to cut fuel costs, and, because the utility has a monopoly, the customers must pay. The pass-through creates an incentive for the ultimate purchaser of energy to cut energy use, and such a response in turn lowers the utility's demand for fuel. But this course of events does not provide the utility with an incentive to make its operating practices more energy efficient. As long as regulatory practice allows utilities to pass on fuel cost increases and requires them to pass on any fuel savings to the customer, their incentive to save on these expenses is severely limited.

The power of utilities to pass on costs to users has occasionally resulted in the discouraging situation of customers paying more for energy because they have used less. As a direct result of decreased demand, some large electric utilities have sought—and received—rate increases to maintain their guaranteed rate of return against fixed capital and labor costs. This can cancel or reverse the financial benefits of energy savings for customers who have made small savings. It is true that these rate increases may be smaller than would have resulted if customers had not cut energy use, because conservation may allow a utility to forgo construction of new and expensive plants. But this point escapes most customers because they generally measure the effects of their conservation activities by their utility bills and not by what a bill would have been without the conservation effort (Kempton and Montgomery, 1982). An economic incentive remains but many utility customers do not perceive it. The utilities are left with a credibility problem, and their customers may give up on any conservation efforts.

As a general rule, organizational energy use is not inefficient, so production processes are not often greatly wasteful of energy. But past procedures are often routinized and may not readily change with economic conditions. Also, when an organization has no incentive to save energy, as with some utilities, or when customers lack the market power to provide an incentive for energy-efficient operation, the problem becomes significant. It has not been considered appropriate for government to intervene in the private sector by regulating industrial processes; for public-sector organizations, however, intervention is acceptable and much more likely.

Providers of Public Goods and Services

Political units make decisions for energy users without necessarily treating them as energy decisions. They may create needs for energy services or provide means of meeting needs at different levels of energy intensity. For example, the political decisions to invest billions in limited-access highways, to offer tax incentives to homeowners, and to develop centralized water and sewer systems, have had major long-range energy implications. These policies have encouraged suburbanization and have made detached suburban homes relatively inexpensive and convenient to reach by automobile. But a dispersed settlement pattern creates a transportation need that is hard to fill except by private automobiles. It also increases heating and cooling needs in comparison with attached city dwellings.

The trend toward suburbanization has further limiting effects on energy choices: as dwellings and workplaces disperse and inner-city populations decline, mass transit systems are used less and therefore become less energy-efficient and more expensive. As a result, transit services are cut, making even more people dependent on the automobile. In this way, past political decisions, such as those to invest in public highways rather than public transit, have significantly changed the environment in which people choose where to live and corporations choose where to build plants and offices. Those decisions have made some choices attractive and effectively foreclosed others, with far-reaching energy implications.

Local political decisions that do not involve expenditures can also indirectly influence energy use: zoning practices, building codes, and land-use plans can influence the placement and construction of buildings in ways than constrain choice both by individuals and organizations. This may be a potent form of influence on energy use, as evidenced by the effects of a passive solar building code in Davis, California (Dietz and Vine, 1982): the code has changed building practices in Davis and some of the surrounding area, and has resulted in a greater saving in energy than could have been achieved by the new construction— so, adoption of the Building code may have changed the behavior of the occupants of existing buildings.

In the past, decisions about building codes, highways, and the like were usually made without energy being considered explicitly. The decisions were governed by the interests of developers, lending institutions, labor unions, industries, prospective home buyers, suburban motorists, and so forth. Energy was cheap and the interests of energy users as a group were rarely represented. It remains to be seen whether these conditions have changed enough over the last decade to significantly alter the politics of local decisions that make subtle but important choices for energy users. It is certainly still the case that energy costs to individuals are not a major concern of those interests that have traditionally shaped local decisions about public goods.

Engineers, Designers, and Architects

The technical experts who create buildings and machines do not always have energy use as their main concern, but their decisions build energy-using characteristics into their products. In the design of refrigerators, for example, considerations of energy efficiency conflict with a desire to produce a product with the greatest possible food storage area given the outside dimensions of the appliance. That is, goals of good engineering conflict with marketing goals. The manufacturer makes the final decision, and in the past often chose marketing goals over engineering goals, leaving engineers to design relatively energy-inefficient products. In building design, an esthetic decision must be made about whether building occupants should be insulated from weather or should experience it. Until fairly recently, architects always made this choice one way, and, as a result, energy inefficiency was built into glass-enclosed buildings.

Costs of energy are, of course, involved in design decisions concerning appliances, buildings, and automobiles. Past decisions were predicated on low costs, and higher costs are a force for change. But other influences also operate within the design and engineering professions. Architecture, for example, has always had vogues, epochs, waves, and so forth. The entry of energy efficiency into architecture can be described in the language of art history: it was embraced initially by an avant-garde (generally young), mocked and repudiated by an establishment, but gradually penetrated the profession. This history ran parallel to rising energy prices, but had a dynamic of its own: the rapid adoption of energy conservation by architects was probably aided by the fact that it was a *new* objective for a profession that values innovation.

Similarly, it is arguable that the history of energy efficiency in automobiles and appliances partly reflects status distinctions within the engineering profession. In the United States, high technology has had most of the status and has attracted a large share of the most creative engineers. The engineers in lower-technology and therefore lower-status jobs, such as those designing refrigerators or automotive drive trains, could not compete with marketers for influence over organizational decisions. As a result, the U.S. public has come to the point where it associates good low-technology engineering (as in cars) with foreign countries. Recent concern about energy conservation may be restoring some dignity and influence to low-technology engineers.

Lending Institutions

Banks and other lenders influence energy use by their willingness to make funds available at favorable rates for energy-efficient improvements in equipment and by their willingness to offer mortgage money for energy-efficient new structures. Bankers' beliefs about the effects of energy efficiency on the quality of mortgage investments are also important. They are more likely to make loans for energy efficiency if they believe it increases resale value, attractiveness to renters, or the ability of the owner to make loan payments. Such considerations have led the Federal Home Loan Mortgage Corporation to consider energy conservation among its criteria for purchasing loans on the secondary market. This criterion may eventually affect the energy efficiency of housing through the behavior of primary lenders.

Standard-Setting Organizations

Standard-setting organizations, which are mainly voluntary in the United States, standardize a variety of energy-related products and services. Many interests, often conflicting, are represented in such organizations. Thus, when the American Society of Heating, Refrigeration, and Air-Conditioning Engineers devises standards for building construction, some organizations represented in the society stand to gain because changed standards will increase demand for their products, while others stand to lose. In the past, the organizations that set standards for heating, cooling, and lighting in buildings usually set standards to increase the comfort of building occupants. Such standards also promoted sales for the organizations whose representatives set the standards. Energy efficiency was not a serious consideration in standard-setting at that time. Now, with energy users more concerned about costs, it may be appropriate to reexamine standards based on comfort to see if they are warranted. These standards have changed greatly in the last few decades: standards today require twice as much lighting in buildings and a narrower band of temperatures in workplaces than a generation ago. The expectations of building occupants have changed at the same time as the standards: many workers expect air-conditioned workplaces in climates where very few had them in 1960.

There is reason to reexamine the basis of existing standards of comfort. Most of the existing research on human response to temperature in buildings has been conducted in environmentally controlled rooms in ways that do not allow for adaptive responses by occupants (e.g., Stolwijk, 1978). Clothing levels, for example, are usually held constant. This research strategy implicitly defines comfort as a physiologically determined function of temperature, clothing levels, rate of air movement, and a few other physical variables. There is evidence, however, that comfort is a preference that people are constantly choosing, rather than a physiological function. One line of evidence is the variability of up to 13 degrees Fahrenheit among average air temperatures in homes in different Western nations (Schipper and Ketoff, 1982). Another is the experimental research that demonstrates that households given a schedule of slow adaptation to lower temperatures in their homes are as comfortable at 62 degrees Fahrenheit after adaptation as at 65 degrees Fahrenheit before (Winett, Hatcher, Fort, Leckliter, Love, Riley, and Fishback, 1982).

As long as standard-setting organizations continue to define comfort as they have, however, the purchasers of buildings are limited in their choices and somewhat constrained in their energy use. They may be purchasing more lighting and larger heating and cooling plants than they need and so using more energy. When standards become incorporated in building codes, building owners and occupants are especially constrained. Thus, there is potential for modifying energy use through changing decisions in standard-setting organizations.

Policy for Energy Intermediaries

It is unrealistic to attempt to offer general policy recommendations for dealing with intermediaries because there is too much variability among sectors of the energy market. As a first step, the importance of intermediaries as possible agents for change should be emphasized. It is worthwhile to compare the impact on energy use that could be achieved through change by the intermediaries involved in a given area of energy consumption with the magnitude of change within the discretion of ultimate energy users.

Policies aimed at actors other than energy users are sometimes highly attractive in terms of potential effectiveness. For example, the potential for energy savings in transportation was much higher from regulating the manufacture of automobiles than from financing mass transit or providing incentives for ride-sharing by travelers (Hirst, 1976). In that particular case, hindsight shows that regulating the manufacturer was politically feasible. The situation has been different in the building industry: the great technical potential for energy savings through more energy-efficient building practices remains largely unrealized, partly due to the political failure to develop a national set of performance standards applicable at the local level. What is technically attractive may or may not be politically realistic, for reasons that are not well understood.

In short, while there is a relatively well-developed body of knowledge that can be applied to developing energy programs for energy users, especially individuals and households, much less is known about how to design and implement policies to change the behavior of manufacturers, lenders, designers, and other intermediaries in the energy field. Part of the problem is that a complex political process is involved in dealing with intermediaries, many of whom are formidable political actors. They can exert great influence when they see an energy proposal, rightly or wrongly, as threatening their interests. There is little reliable knowledge about how to resolve political impasses caused by a divergence of interests between a group of energy users and a group of producers or intermediaries. Sometimes a manufacturer may become convinced to reassess its interests, as the automobile industry seems to have done in the case of fuel-efficient cars. Sometimes energy users' interests may be successfully mobilized against the interests of an intermediary group, or the balance of forces among affected intermediaries may be such as to promote an energy policy to which one major interest is opposed. But there is no research knowledge that enables predictions of which of these outcomes is likely.

Given the limited available knowledge, we emphasize a few relatively simple points. First, intermediaries are often at least as important as energy users for influencing energy use. Policy makers and the public would do well to look carefully at intermediaries as foci for action.

Second, intermediaries are political actors with some power to promote or interfere with the implementation of policies that affect them. For those who see a national interest in energy efficiency, there is great potential support as well as strong opposition from intermediaries—finding allies may be politically critical. It should be remembered that intermediaries sometimes reassess their interests. For example, the idea of performance standards or goals for buildings may at first strike architects or builders as unwarranted regulatory interference. But when there is low demand in the market for new construction, a performance standard may look more attractive: it may create a demand for design services and set a target for upgrading old buildings. Because of the special risks associated with energy-efficient innovations for some private-sector intermediaries, they have a disincentive to invest in national goals unless there is a return available for them. Therefore, if a public interest in energy efficiency is accepted, attention must be given to overcoming barriers to change among intermediaries.

Third, intermediaries that innovate in energy efficiency have a stake in educating their potential customers to the value of their innovations—

especially those innovations that are invisible to customers and impossible to monitor until after purchase. But intermediaries are rarely skilled educators, and in any event have a conflict of interest when it comes to educating customers about their products. This situation suggests a role for the public sector in educating purchasers of buildings, machinery, vehicles, and other energy-consuming stock about the implications of energy-related innovations for their interests.

Fourth, most intermediaries are organizations, and principles of organizational behavior apply. For example, intermediaries are apt to make decisions by following rules. Changes in organizational procedures to increase attention to energy should make a difference. An example is the practice of the Federal Home Loan Mortgage Corporation of considering energy efficiency before purchasing mortgages.

Two recent innovations in energy efficiency through intermediaries are particularly worthy of further study, development, and dissemination. Experiments should be done to negotiate agreements to share the costs and benefits of energy-efficiency investments in buildings between owners and occupants. Because arrangements that work in one location are not automatically transferable, we recommend that such pilot projects involve interested parties outside the particular building, such as consumer groups and business associations. This may help spread the word of a successful agreement and also commit the interested observers to trying to adapt any workable agreement to other situations.

Evaluations should be done on the effects of existing programs to certify the energy efficiency of multifamily buildings. If these programs show promise, increased efforts to gather performance data on buildings are warranted, so that certification programs can be easily expanded.

Notes

1. Organizations in a relatively unfavorable environment are more likely to engage in legally questionable activities (Lane, 1953; Staw and Szwajkowski, 1975); university departments introduce more client-related innovations in courses under conditions of relative adversity (Manns and March, 1978); relatively unsuccessful firms take the lead in the adoption of some specific, extremely important technological improvements (Adams and Dirlam, 1966). Studies of research and development activities in firms similarly suggest that failure may stimulate research and development (Williamson, 1964; Schott and von Grebner, 1974; Kay, 1979).

2. Mansfield's (1961) study of the introduction of twelve different innovations in four industries did not support the "innovation in the face of adversity" hypothesis. His subsequent work (Mansfield,

1963, 1964) indicates more support for the idea that large, successful organizations tend to innovate sooner than do smaller, less successful organizations. Daft and Becker (1978) in their empirical study of the adoption of innovations in thirteen high school districts obtained results that do not support the notion that failing organizations innovate more than successful organizations. Oskamp (1981) found that larger chemical companies introduced energy conservation earlier than did smaller firms, and Mills (1981) found that large retailing chains had engaged in more energy-management activities than had smaller chains.

3. Rules, routines, and standard operating procedures determine the pace of decisions (Fischer and Crecine, 1980), information flows in decision making (Wohlstetter, 1962; Allison, 1971; Pettigrew, 1972), and the ways in which inputs are combined to make decisions (Crecine, 1967; Allison, 1971; Britan, 1979).

4. For example, change may be the result of crises (Chandler, 1962; Hall, 1976; Johnson, 1978), entry into new product lines (Chandler, 1962), changes in environmental conditions (Ritchey, 1981), or shifts in management ideology (Schiff, 1966). Routines may also change informally as a result of managerial judgment about their appropriateness or efficacy in a given situation (Sproull, 1981b), subordinate perceptions of the expectations of superiors (Ellsberg, 1972), or as a hedge against uncertainty (Larkey, 1979).

5. Williamson (1975) and others have related organizational forms to the problems of forming contracts to manage uncertainty. Pfeffer (1972a) found that Israeli managers who were dependent on the government for sales or financing tended to be more complacent about governmental policies than other managers. Salancik (1979) found a similar result in the United States, though it was affected by the public visibility of the firms involved. It has been argued that intermediate levels of concentration in an industry are distinguished by high levels of uncertainty and have been shown to be associated with the frequency of mergers (Stern and Morgenroth, 1968; Pfeffer, 1972b; Pfeffer and Leblebici, 1973), joint ventures (Pfeffer and Nowak, 1976), interlocking directorates (Pfeffer and Salancik, 1978:166; but see Palmer, 1980), and executive movement across organizations (Pfeffer and Leblebici, 1973).

6. Imitation should imply identifiable innovation leaders within an industry, and Webster (1971) reported some consistency across innovations in the airline industry. However, most studies find that firms that adopt one innovation relatively early are not nec-

essarily early adopters of another (Mansfield, 1968; Webster, 1971).

7. This information comes from L. McClelland, Institute for Behavioral Science, University of Colorado, Boulder, 1982.

8. See note 7 above.

6

Energy Emergencies

One of the most dramatic lessons from recent history is that the system of energy supply and use in the United States is vulnerable to disruption. When oil represents nearly half of the energy used and imports account for a large portion of that oil, the country is susceptible to actions that would sharply reduce the amount of oil it can import. Moreover, recent experience shows that electricity blackouts, coal strikes, and natural gas shortages in winter can and do occur. And sudden, large-scale energy shortages can dramatically increase energy and other prices, cause serious unemployment, and lead to social and political stress. In short, the prospect of a serious energy shortage underscores the views of energy as social necessity and a set of strategic materials (see Chapter 2).

Because of the countless ways in which energy has become essential for comfort, convenience, mobility, and productive activity in the United States, few events seem potentially more disruptive than a sharp reduction in its supply. Most people agree that the United States should try to reduce the threat posed by any or all of these energy contingencies. One explicit indication of this has been the decision—with remarkably little controversy—to create a national strategic petroleum reserve, a public investment that could eventually cost more than $30 billion.

For the past ten years, much of the analysis of energy disruptions has taken either of two approaches: the first has been purely technological, focusing on the capacity of the energy production and distribution system to deal with disturbances; the second has emphasized the relationship between public policy and market adjustment processes. The latter approach has used the theory of market allocation as a normative standard and analyzed the effects of various government policies, such as price and allocation controls, on the adaptive responses of the economy.

This chapter takes a different view of the emergency process, emphasizing what we call its human dimension—the behavior of individuals and institutions in anticipation of energy emergencies and the effects of an energy supply interruption on social institutions, social processes, and individual well-being. This perspective recognizes that despite the theory of market allocation, governments have repeatedly intervened in emergencies. Various social goals and social processes, underemphasized in market analyses, play a major causal role in generating political demands for government intervention and are also important in their own right. The best example is probably the fact that energy emergencies are perceived as human failures. This perception leads to efforts to place blame, and political conflict over responsibility predictably follows.

THE VARIETY OF ENERGY EMERGENCIES

We define an energy emergency (after Fritz, 1961) as an event, concentrated in time and place, in which an interruption in energy supply or use disrupts essential functions of a society or economy.

Two aspects of this definition are worth noting. First, an emergency is a temporary phenomenon. Emergency preparedness is distinctive in that it concerns preparations for and responses to short-run, abnormal conditions. Second, an emergency exists because the economy and society are disrupted, not just because energy supply or use is interrupted. Energy is a means to social ends; it is not an end in itself. When those social ends can be achieved with substantially less energy, even on relatively short notice—by using stored energy, reducing nonessential uses of energy, substituting other things for energy, or changing people's views about comfort and convenience—the tie between energy events and true emergencies is weak. At the heart of energy emergency preparedness, therefore, is the relationship between energy and society.

Dimensions of Energy Emergencies

Energy emergencies differ from each other in important ways, and those differences affect the role government needs to play and the kinds of policies it will undertake. At least eight dimensions of energy emergencies are important:

1. Causation.   The cause of an emergency may be internal or external to the affected area: for example, a power plant breakdown or a foreign oil embargo. It may be a conscious goal of policy or an unintended side effect. The precipitating event may be political, economic, military, legal, or even natural.

2. Immediacy.   The onset of an emergency may be immediate with respect to the energy event that causes it, or the effects may be delayed.

3. Magnitude.   The effects associated with the emergency may range from discomfort to catastrophe.

4. Incidence.   The emergency may be concentrated in a small area, or it may be diffuse. The disruptive effects of an emergency may be focused on one segment of society or may be defined by geography, income level, or energy dependence.

5. Duration.   The emergency may be brief or long-lasting. As in the oil crisis of the 1970s, it may be the acute phase of a long-term socioeconomic change.

6. Probability.   An emergency may be considered likely or unlikely.

7. Credibility.   In some emergencies, official statements that an emergency exists are little questioned, for example, a widespread power failure that leaves a large group without light, heat, and power; in others, such statements result in public skepticism, for example, an announced shortage of oil reserves.

8. Manageability.   An emergency may be one whose likelihood or effects can be easily reduced, or it may be exceedingly hard to handle.

Clearly, emergency preparedness is shaped by the kind of emergencies one considers. Low-probability events present different challenges from high-probability events, and a situation in which social disruption would be immediate is fundamentally different from one in which the disruptive effects of the event lag by weeks or months.

Types of Energy Emergencies

Table 2 illustrates the variety of energy emergencies by comparing four types of emergencies. As it shows, an oil import emergency is in many ways unlike a natural disaster or other kinds of events. For instance, an emergency because of an interruption in oil exports from the Middle East is distinctive in the time lag between the drop in exports and the actual drop of energy supply in the United States. It also differs from most other kinds of energy emergencies in its relationship to international politics, its widespread impact throughout the country, its particular focus on the transportation sector, and the process through which it becomes apparent. Past experience suggests that an oil emergency may also be associated with particularly serious credibility problems for large energy corporations and the federal government.

An electricity emergency due to a shutdown of nuclear power generation, as shown in Table 2, is distinctive in that it results from a conscious regulatory decision. It differs from an oil import emergency in the immediacy of its impacts, as well as in its particular focus on regions and sectors.

A coal strike resembles an oil import emergency in that, because of coal stockpiling by major users, energy shortages are not immediate. It differs from an import emergency, though, in the influence the U.S. government can have on the causes, the timing, and the duration. Its focus on coal-producing and coal-using regions and sectors is different from either an oil shortage or a nuclear electricity curtailment.

An "act of God" is distinctive in its cause and usually in its unpredictability, its immediacy, and its strong regional focus. It may also differ from other kinds of emergencies because people are less likely to assign blame or seek scapegoats for it; consequently, it may be less divisive and more likely to generate social cohesion than other types of energy emergencies.

Other distinctive sets of conditions can also be imagined. For example, less immediate "acts of God" can also lead to energy emergencies: droughts as they affect hydroelectric power supply and protracted cold or hot weather as it affects energy demand in buildings.

This chapter is concerned with serious national emergencies, including oil and coal supply disruptions or a nationwide nuclear power plant shutdown following a local nuclear plant catastrophe; it does not discuss emergencies of small magnitude or those whose effects are very localized. It is necessary to take the possibility of a major energy emergency seriously, even though U.S. history since World War II has left Americans with the expectation that any emergencies that may arise will be of only moderate intensity. (Views about nuclear war are an obvious exception to this.) This expectation, if not held as matter of conscious and considered belief, has been a tacitly accepted context for thought and discussion about emergency preparedness. Associated with the general expectation that emergencies will not be severe are some more specific expectations about the way society will respond. These expectations are also historically based. For example, people tend to expect that "business as usual" will continue to a large extent in spite of an energy emergency. Consequently, it can be expected, and has so far been the case, that the political struggles of organized groups to secure economic advantages through the political process will continue unabated, and may even be intensified, in an energy emergency.

Table 2. Dimensions and consequences of selected energy emergencies

Dimension	Type of Emergency			
	A major interruption of oil supply from the Middle East	A shutdown of nuclear electricity generation because of a nuclear accident	A protracted coal strike	An earthquake in the south-central U.S. that seriously damages major oil and gas pipelines, rail lines, and electricity lines in the region
1. Causation	External governmental decisions, political instability, or terrorism	Regulator decision to protect public safety	Labor dispute	Natural hazard ("act of God")
2. Immediacy	Import cutoffs lagged 5-6 weeks	Immediate effects on electricity supply	End-use energy scarcity lagged up to 90 days	Immediate effects on gas supply to importing regions, and on the electricity grid; various lags otherwise
3. Magnitude	Uncertain; could be serious but unlikely to be catastrophic	Could be substantial in some locales	Probably no more than discomfort in most places; union control over coal supply appears limited	Possibility of immediate dangers to life and health in the affected area but energy impacts unlikely to be the major concerns locally
4. Incidence	Affects most sectors of society and economy in the U.S. and in	Some geographical focus but broad social effects in region where	Focused on coal-producing and coal-using regions	Besides the earthquake region, focused on other regions

	other oil-importing countries; particular impacts on transportation and other end-users of oil (heavy industry, some utilities, building in some regions)	nuclear power is important, and on the nuclear industry	and sectors	receiving energy shipments through the south-central U.S.
5. Duration	Uncertain; unlikely to be very short, could be quite long	Uncertain and probably contentious; likely to be long	Strike must be fairly long to create an emergency	Short—until repairs are made and shipments resumed
6. Probability	Opinions differ, but may be fairly high	Opinions differ; probably low, but with a rising probability if/as nuclear power generation grows	Low to moderate; probably varies with national dependence on coal	Low
7. Credibility	Uncertain, unless the cause and magnitude of the emergency are self-evident; considerable public distrust of some of the key institutions	Probably high, but disputes about the need for the energy action would grow as impacts are felt	Higher in coal-producing than coal-using area	High
8. Manageability	Causation is hard to control, but the impacts on an import cutoff can be reduced by prior actions: stockpiling, fuel shifts, oil conservation, and domestic energy production	There are ways to reduce the likelihood of the emergency, but the impacts of an emergency, if it occurs, would be hard to handle	The relative lack of immediacy makes this situation more manageable than most	Almost impossible to prepare for or to manage in the very short run; rapid recovery is practicable

But there is no basis for great confidence that serious emergencies will not occur. It is all too easy to generate plausible scenarios by which a severe emergency—for example, a virtually complete cutoff of Middle Eastern oil supplies—might come about. The sequence of hypothetical headlines as such a scenario unfolds is no more bizarre than some actual sequences that have occurred in the last few years. To restrict analysis and policy to contingencies of moderate severity, like those of recent decades, is no more reasonable than ignoring the possibility of emergencies entirely.

This chapter sets forth some insights from social and behavioral research on emergencies that may help to clarify some problems and point the way to some solutions. Some of the generalizations are robust enough to apply across a wide range of energy emergency situations; others apply more specifically to a particular kind of crisis situation, such as an oil import emergency.

PREVENTION AND PREPAREDNESS

Being prepared for an emergency can sometimes help in preventing it, but preparedness is no substitute for prevention. Preparedness is only likely to be efficient and credible if it is coupled with reasonable efforts to prevent energy emergencies.

The attitudes of the public toward the federal government and other large institutions during an emergency—a key to an effective response—depends partly on whether or not people believe that everything reasonable has been done to avoid the emergency. Credibility in this respect is a key to effective response because it is known that some parties will profit in certain emergencies and it is widely suspected that those parties do not want to avert an emergency. For example, most people in the United States believed that the 1973–1974 oil crisis was caused by the government or by the oil companies (Murray, Minor, Bradburn, Cotterman, Frankel, and Pisarsky, 1974), and there remains a common belief that oil and natural gas have been withheld from the market to force price increases (see, e.g., Brunner and Vivian, 1979). In order to combat cynicism and to encourage cooperation, a comprehensive approach to emergency preparedness must have as its fundamental thrust the anticipation and prevention of sudden changes in energy supply.

In fact, prevention and preparedness are part of the same process. For example, one approach to dealing with an oil import emergency is to prepare to deal with certain specific effects of a supply interruption: preparedness, in a narrow sense. Alternatively, the nation might try to prevent the kinds of international crises that might precipitate an interruption or to reduce its dependence on oil by diversifying its energy sources.

Clearly, these different kinds of approaches are not independent of each other. Being prepared for an emergency may help to prevent it. For example, several types of emergencies can be caused by a party who wants to use their effects, or the threat of those effects, to achieve an objective. If the action—limiting exports, blowing up a power plant or transmission facility, going on strike, or whatever—will not have serious economic and social impacts, it loses much of its purpose and potential power. Other possible causes of emergencies will remain, but the overall probability of a crisis will have been reduced.

At the same time, prevention can aid preparedness by making it more credible. Everyone knows that there are winners and losers in almost any set of circumstances, and that some, including parties in government or in industries, may see emergencies as ways to solve other problems. For example, an emergency can be used to establish controversial policies. Given these realities, together with the natural tendency in unstable times to look for conspiracies and to assign blame, any impression that the country has not tried hard enough to prevent a crisis could rapidly turn into a serious alienation of broad segments of the population under the stress of an emergency.

One of the primary difficulties in establishing consensus about energy emergency policies is that different strategies look more or less appropriate depending on the observer's time horizon. With a short time horizon, it makes some sense to plan as if emergencies will not occur, but with a long horizon, emergencies are a certainty, and prevention becomes a critical goal. For example, major manufacturers in the United States require fairly long-term stability in energy availability in order to make judgments about production, markets, and investments. Such stability can only exist when there is a sustained attention to what some might call "system maintenance" with respect to energy conditions. A long time horizon is at the heart of such policies.

An important element of prevention is resiliency: the social and economic capacity to adapt to uncertain domestic and international energy supply and demand conditions. Because resiliency, or the lack of it, is not evident until a system is stressed, it is difficult to conceptualize as a policy goal. It does, however, sometimes conflict with short-term efficiency goals: under short-term decision criteria, investments in resiliency may appear redundant or economically inefficient. Consider the problems of a potential producer or consumer investment in an alternate energy system that relies on an unconventional source of fuel. As long as the primary fuel is available at an affordable cost, conventional corporate accounting would treat the proposed investment in a new system as unproductive, since it in fact produces no concrete goods or services. But the resiliency of a firm would be increased because an alternative or a backup system would be productive when the usual fuel is scarce. The presence of such a system might have an additional value: it would change the executives' decision-making environment, permitting capital and labor allocations with less uncertainty about the future availability of energy (see Chapter 5).

The simplest parallel for investments in energy resiliency is risk or hazard insurance. No enterprise or individual in a modern industrial society can operate without some form of such private or collective insurance in a variety of areas: crop insurance, flood insurance, nuclear power plant insurance, foreign investment insurance, and so forth. The principle of risk insurance has been accepted as a part of the transaction costs of a modern industrial system, and except when the insurance fails to cover a loss, there is no tendency to treat it as an unproductive investment. Rather, models of the cost-effectiveness of insurance have been developed, and such cost-effectiveness models could be extended to resiliency investments as well. The fact that there is no simple technical measure of "enough resiliency" should not impede recognition of the basic need for some resiliency in a complex industrial system—not a social or political consensus about a single mode of energy resiliency, but rather an evolving agreement about more or less reasonable proposals and investments, to be altered as political and energy conditions change.

A long time horizon becomes intertwined with resiliency when one makes an explicit effort to reduce the likelihood of energy emergencies. For example, preparing for a possible oil import cutoff might include improving mass transit capabilities, thus diversifying the transportation system. Prevention policies might include an emphasis on fuel-switching capability, so that a home or office building could run on oil, gas, solar, or geothermal energy. Prevention might imply energy policy and geopolitical agendas that reduce the strategic importance of the Middle East and reduce dependence on energy and other imports. Prevention might also be implicit in a policy stressing further development of service industries rather than energy-intensive product industries.

Prevention policies would ensure that the country does not have all its energy eggs in one basket. Policies and actions that reduce reliance on one or a very few specific energy options or facilities would be preferred, because diversification—of resources, technologies, and sites—is one of the most important protections against energy emergencies (see, e.g., Rochlin, 1977). In this sense, strategic oil storage is a preventive measure because it reduces reliance on particular energy suppliers. Improved energy efficiency is a prevention strategy for another reason—it stretches available supplies. Even the design of buildings and equipment can be part of a prevention strategy. For instance, in the case of a summertime power outage, consider the adaptability of a high-rise building with windows that open compared with one whose windows do not.

Prevention extends to the vulnerability of other countries as well as the United States. A truly significant reduction in world oil supplies would affect the United States even if it were not directly dependent on oil imports, because world economies are intricately linked. Economic disruptions in Europe and Japan would undermine the U.S. economy and national security as well, although the impact would be less immediate. Consequently, policies that reduce oil dependency for Europe, Japan, and other oil-importing nations will also reduce U.S. vulnerability to emergencies.

Some of these preventive actions can be taken by reallocating existing resources—focusing design efforts on promoting energy efficiency in buildings, for example. Many forms of prevention, however, require resources—money, skills, and the attention of corporate and government leaders—that might otherwise be allocated to emergency preparedness or other objectives. Prevention can reduce the cost of preparedness by reducing the number of contingencies being prepared for. Thorough preparedness is costly, not only in direct budget requirements but also in public attitudes, which in turn affect preparedness and emergency response. However, public investment in preparations that help very little, or are perceived to help very little, in an actual emergency can exacerbate problems of confidence and credibility. A process that cries "wolf" too often can desensitize people to threats in the long run. Prevention aids preparedness by helping it to be credible and by enabling it to focus on the emergency conditions that are hardest to prevent.

DIVERSITY AND CONFLICT IN ENERGY EMERGENCIES

There are vast differences within an affected population in attitudes toward energy emergency preparedness and in needs and responses during an emergency. Preparations should be sensitive to this diversity and, when possible, make use of it as a part of an effective response to an emergency.

The American population is a collection of very diverse groups with respect to energy emergency preparation and response. Some individuals and groups will perceive a particular situation to be an emergency when others in similar circumstances do not. Seeing an emergency, different people will react differently: some will accept official pronouncements about the emergency; others will not. In addition, when people accept the reality of an emergency, their options are often constrained: for example, what makes sense for young people may be unrealistic for the elderly, and some options available to the wealthy will be unavailable for the poor, as noted under "Incidence" in Table 2. These differences will create a diversity of behavior that contains the seeds of social discord and presents a formidable challenge for emergency management.

The Importance of Decentralized Response

One clear implication of diversity is that, especially in diffuse emergencies, detailed central control of emergency responses is impossible. Many decisions will require a knowledge of local attitudes, capabilities, and needs as changes in patterns of supply and use create a mosaic of localized gluts and shortages and as communities move to deal with their own problems. Because a central authority simply cannot manage the information flow and integrate the many pertinent factors with the speed and sensitivity that would be required, much of the troubleshooting in an emergency will need to be decentralized.[1]

The market system is one important mechanism for this decentralization. But it is important to recognize that social groupings, linkages, and stresses may be equally important during a serious emergency. Studies of communities under disaster conditions indicate that family groups, social relationships, and community organizations are often the focus of behavior (Dynes, 1970, 1972). In an emergency in which transportation fuel becomes painfully scarce, for example, friendship ties in neighborhoods and workplaces will influence the formation of carpools, and local nongovernmental organizations may take the lead in ensuring that community needs are met. Furthermore, attitudes and responses in a crisis situation are strongly affected by social interaction, as well as by the prices of goods and services. Individual opinions as to whether or not an emergency actually exists seem to be influenced by the behavior of close associates (Latané and Nida, 1981); emergency preparedness may also be influenced by social contacts and groupings. In a sense, it is a kind of social innovation, spreading through contact networks, which may be accepted or rejected. It is shaped by the salience of preparedness responses as alternatives and by the examples of others (e.g., Staub, 1974; Wilson, 1976), as well as by the norms of groups with which one is identified (Buss, 1980; Schwartz and Clausen, 1970).

Because of such phenomena, an effective approach to emergency preparedness must address itself to the diversity of local concerns. It may, in fact, even be able to take advantage of some of this diversity, both in preparing for an emergency and in preventing it. While centrally located officials are often unable to deal with rapid changes in attitudes, behavior, and structures in an emergency, spontaneous improvisation at the local level has often been a key to effective emergency response (Kreps, 1979; Quarantelli and Dynes, 1977; Barton, 1969). Decentralized initiatives come from local groups that sometimes represent neither conventional economic actors nor conventional government operations. An example is the role of automobile clubs during recent oil shortages: by providing accurate information about the local availability of gasoline, the clubs helped to reduce disruption by adjusting travel to energy availability.

Actions can be taken ahead of time to facilitate this sort of appropriate decentralized social response. For example, plans can be made for using school buses during off hours for public transportation, especially if certain issues, such as arrangements for drivers' overtime, insurance, and the method of collecting fares, can be resolved beforehand. The solutions reached in one community can be shared with others through professional associations. Emergency preparations should link a wide variety of local groups and agencies, so that the capability for decentralized response can flourish and spread.

Finally, local differences can be a key to reducing the likelihood of emergencies. For instance, alternatives to imported oil as a source of energy often look more attractive when they are evaluated at a local scale (Lovins and Lovins, 1981; Wilbanks, 1982).

The Likelihood of Social Conflict

Although social cohesion often increases after a natural disaster, energy emergencies are likely to be socially divisive. Because most energy emergencies are perceived as human failures, placing blame, political and legal conflict, and mistrust may be widespread.

Most energy emergencies are seen as failures of a technology, a policy, or an economic or governmental system. This judgment leads people to identify individuals, organizations, or institutions as responsible for the disruption. Since the causes are usually complex, people will almost certainly disagree about blame, but a few tendencies can be anticipated. In hindsight, the emergency will seem to have been foreseeable (Fischhoff, 1975a, 1975b; Fischhoff and Beyth, 1975). Observers will think that the responsible party should have acted to prevent the emergency or at least to prepare for it. Furthermore, people are likely to attribute the behavior of others to personal motivations rather than situational constraints (e.g., Jones and Nisbett, 1971; Ross, 1977), and whoever is seen as responsible is more likely to be labeled stupid, incompetent, or malevolent than unfortunate, uninformed, or confused. The responsible party often loses credibility, sometimes irreversibly. Because blame is likely to be spread widely, at least for a large-scale emergency, this reaction endangers the reputations of many parties who may have been able to do little to prevent the emergency, at least in the immediate past.

Such a situation is less likely to arise if the cause of the emergency is seen as outside the U.S. economy or society rather than within it. Generally, when an event provides a social group with a common adversary or a common goal, a strong and cooperative response is likely (e.g., Dion, 1979; Goldman, Stockbauer, and McAuliffe, 1977; Sherif, Harvey, White, Hood, and Sherif, 1961); wartime crises and natural disasters are examples. But when an event is viewed as rising from within—due, for example, to a failure of leadership—social turbulence usually occurs. Thus, high unemployment often leads to political change. The widespread suspicion that seems to have accompanied recent energy crises in the United States suggests that oil shortages fit in this latter category.

Differing economic self-interests can add to the strains of an energy emergency. Some individuals and groups will see each other as rivals for scarce resources—the violence in gasoline lines is an example—and groups that gain in the crisis will probably receive blame for the emergency itself or for the area's lack of preparedness for it. And other conflicts will arise. For example, in an oil shortage, a firm may find it more attractive to shut down for the duration than to install an alternative heating system that will not be needed when the emergency is over. To employees and customers, however, this may seem to be an unacceptably disruptive strategy. Some categories of energy users may appeal to government for protection from price or quantity impacts of an emergency, but the more categories that are protected, the more severe will be the impacts on the remainder. Giving agricultural users or car rental firms extra gasoline allocations, as in the recent federal allocation plan, reduces the gasoline available for other users who believe they are equally, or more, deserving.

Additional sources of friction can be found in limitations inherent in economic markets. For example, an emergency is likely to lead energy users to seek substitutes for scarce commodities—emergency housing or warm clothing for heating fuel, wood for fuel oil, electricity for natural gas, and so forth. Because of lags in market responses to such changes in demand which can take shape rapidly as people rush to copy the responses of others, sudden price increases and further scarcities could occur, aggravating concerns about equity and increasing the tendency to look for scapegoats.

Conflict may also arise from sharp and seemingly mysterious fluctuations in energy conditions. For instance, although oil storage is a stabilizing factor in the event of an import cutoff, unused storage capacity is destabilizing, because it encourages rapid private stockpiling if there is the threat of an emergency. If the unused capacity is substantial, prices could rise very rapidly as supplies available for sale disappear. This could occur even when enough oil is available at the start of a crisis to meet needs for use without a large price increase. When storage capacity is filled, the crisis may appear to ease; private stockpilers would be inclined to sell oil from inventories, causing prices to drop. A further threat could send prices back up again. In this way, a period of uncertainty could create a repeated "boom and bust" pattern in the oil supply system (see Forrester, 1961), raising questions about the motives of key parties and creating confusion and distrust among energy users.

PREPARING FOR SURPRISES

Because of the diversity of energy emergencies (as shown in Table 2), it would be a mistake to focus energy emergency planning on a single kind of emergency or a particular "scenario" of possible future events. Even the most thorough preparedness for one type of emergency does not ensure preparedness for a different type of emergency or for energy emergencies in general. And even if it were good to focus on a single type of crisis, the available analytical tools and approaches would not be a reliable basis for anticipating future crises (Kahneman, Slovic, and Tversky, 1982). These tools and approaches are intended for familiar situations, gradual changes, and likely events rather than for extreme cases or unlikely events (Boulding, 1979). The actual problems created by an emergency are often very different from those expected by planners (Fritz, 1967); consequently, the relevance of most modeling research to crisis decision making is limited (e.g., Brewer, 1973; Dutton and Kraemer, 1980; Lee, 1973).

Linkages

Another reason for broadening the base of emergency preparedness is that, although possible emergencies are diverse in their character, they can be linked in their causation. During a winter oil shortage, many people who normally heat buildings with fuel oil will look for ways to switch to electricity: they will often use resistance heaters in each room. Increases in electricity demand, which would be most pronounced in peak-use periods, would hit electric utilities at the same time that oil might be needed for operating peak-load electricity generation units, and the overload could cause regional electricity brownouts or blackouts. As another example, if coal became a critical energy source during an oil emergency, unions unhappy with wage levels or working conditions could use the emergency as a lever for trying to resolve long-standing disputes in a favorable way.

In addition to these horizontal linkages between possible emergencies—for example, between petroleum and electricity or coal—there are also longitudinal linkages in the sense that an emergency at one time can affect preparedness at a later time. If strategic petroleum reserves are used in responding to a particular crisis, they are unavailable to meet a second one until they are replaced. Thus, recovery is an important aspect of emergency preparedness.

The diversity and linkages of energy emergencies make it unlikely that any given set of plans will be appropriate for the next crisis. Consequently, energy emergency preparedness should generally emphasize functions rather than products and facilities; flexibility rather than optimization; and resiliency under a wide range of conditions rather than efficiency.

Preparedness as a Continuing Process

Because of diversity and linkages, energy emergency preparedness should be structured as a continuing process, not as an episodic prepackaging of intended responses. Experience shows us that emergency plans prepared well ahead of time seldom get used when the emergency arrives. Because what actually happens cannot be precisely forecast, a plan may anticipate a situation so different from the actual one that it is of limited value. Moreover, current decision makers may be unfamiliar with plans prepared in the past, may lack confidence in them, and may prefer to rely on their own advisers and ideas rather than the products of people who are no longer in positions of responsibility, and who may be associated with opposing points of view.

If preparedness relies on advance plans for specific events, decision makers are likely to be unprepared when an emergency occurs. One way to prevent this eventuality is to view energy emergency preparedness as a continuing process and to focus on systems for response rather than on specific responses: for example, on systems for communicating during an emergency rather than prepackaged batches of information, as discussed below. Such an approach is especially important if there is a likelihood of a series of emergencies over time. A flexible structure, emphasizing decentralized problem-solving and ensuring that linkages are realistic and manageable, has the potential to cope with a wide variety of situations— even if they arrive in rapid succession. Furthermore, the structure can be modified and adapted to respond to experience, adopting strategies and practices that have been found to work well and changing those that have not.

Some emergency preparation should be aimed at making it possible to respond to a range of unforeseen events. Just as an expedition into unknown territory takes extra supplies along, as well as planning to acquire some along the way, a prudent policy maker should take some preparatory actions to minimize fumbling and delay if an energy emergency hits. As one example, during a serious oil shortage, knowing the location and capacities of gasoline storage tanks at retail stations might become important. To track this information routinely for perhaps 120,000 retail establishments, some of which are always going into or out of business, would be expensive, burdensome, and possibly unreliable. For emergency preparedness, the requirement should be to find out who knows these things as a matter of course, so that they can be queried if necessary. In many communities, for instance, the fire department registers such tanks. The general strategy is discussed further on page 150. Orienting preparation toward the right targets and asking the right questions—such as who can tell us about private gasoline storage during an emergency, rather than how much gasoline is in private storage now—is a part of the answer.

A major problem with emergency preparedness as a social activity is that an interest in preparedness is notoriously hard to sustain. When conditions are normal, people and institutions tend to be preoccupied with more pressing matters than possible future crises, particularly if no emergencies of that type have occurred recently. This is true of both the general population and of people and institutions with crisis management responsibilities. The problem may be especially difficult for energy emergencies because previous experiences have not been serious enough to justify preparations ahead of time. In addition, some people see the spectre of energy emergencies as a devious strategy to justify actions by big government and big business. The possibility of oil import interruptions is often dismissed because people believe that there would be enough lead time to make needed arrangements between the time exports were interrupted and the time the shortfall occurred.

This is not a simple problem to solve, but one part of the solution lies in the value of emergency planning as a process, apart from the plans that result. Especially if preparedness is to be largely decentralized, it would help to involve many people from a cross-section of local society in the process. A continuing effort to develop and test emergency plans can stimulate thinking, alert people to problems and responsibilities, and help to train them in the use of communication systems and other important skills. There is some evidence that practicing how to deal with disruptions may give key people a greater feeling of competence in dealing with special circumstances, even if an actual crisis is different from what has been practiced (Dyer and Fiske, 1982; Kiesler and Sproull, 1982; Thompson, 1981). At least, practice gives key actors experience communicating with each other.

Another part of the solution may be to learn more about how the norms of groups affect the behavior of individuals. If various groups—such as church groups, civic clubs, and public interest groups—decide to become involved in energy emergency preparedness, they can do a great deal to motivate individuals and other institutions. And many of these kinds of groups respond to leadership at the national level.

Still another important factor is the role of the mass media: newspapers, magazines, movies, radio, and television (Turner, 1980). A potentially powerful contribution of the media is simply that of giving attention to the problem of energy emergency. Such attention would help create an atmosphere in which public figures would also pay attention to emergency prevention and preparedness. Emergency preparedness does not necessarily have a very large political constituency (e.g., Wright et al., 1979), but policy makers take many of their cues from the media, which both stimulate and reflect such constituencies.

INFORMATION NEEDS IN EMERGENCIES

Behavior during an energy emergency will be strongly affected by the availability and credibility of information about emergency conditions and options. During an emergency, people and organizations are more information dependent than usual (see, e.g., Wilder, 1978; Newtson and Rinder, 1979). Under normal conditions a person contemplating a vacation trip by automobile needs little or no detailed information about gasoline availability; availability is assumed. But during an oil import emergency, such information may be the determining factor in deciding whether to make the trip. Emergencies require different kinds of information, often sharply focused on details and local circumstances (Newtson, 1973). In an emergency, information becomes a more central concern for organizations that produce it.[2]

Thus, information systems designed for normal situations are usually inadequate in emergencies. They may become congested and break down (Anderson, 1969); they may lack detailed information about regional and local energy supplies; and they may be unable to resolve public concerns, for example, to explain the behavior of large energy corporations. In addition, the information system itself may be disrupted by an emergency, because information gathering itself has energy requirements. Such inadequacies are very difficult to correct on short notice.

When information flow is slow or incomplete, further problems often result. Everyday processes of information exchange and influence will tend to fill any gaps that appear, and thus rumors and misinformation will result. Because many responses are to information about the emergency, rather than to the emergency itself, this problem can become a major factor in emergency management. Another problem is that people are seldom aware of all the possible emergency response strategies available to them (Burton, Kates, and White, 1978) and act within the bounds of the alternatives they can identify (March and Simon, 1958; Slovic, Kunreuther, and White, 1974). There is considerable evidence that when people must rely on a single means of coping with an emergency, they are less adaptive, and the adverse impacts of the emergency on them are greater (White, 1964). A key to reducing the confusion and hardship associated with an emergency is ensuring that people have full knowledge of their options. These concerns imply a crucial role for government in emergency information systems because it is hard to imagine that private institutions will be willing and able to gather and distribute all the information that will be needed in a crisis.

Problems of Credibility

Government faces a serious problem in the credibility of its emergency information. Because many people have limited confidence in government (e.g., Brunner and Vivian, 1979), it would be difficult to assure that government-supplied information will be trusted, even if it is completely correct. The credibility problem is magnified by the fact that government relies on the energy industry for much of its information on energy supply. A fundamental element of emergency preparedness, therefore, is determining how to communicate credibly, reducing unfounded rumor and suspicion, and finding ways for the public to gain access to useful information even when particular messages may consist of unverifiable statements about the future.

For instance, assertions that an emergency is at hand are not always heeded; many people deny danger in emergency situations (Fritz, 1967). They tend to look for reasons to be optimistic, to wait for corroboration from other sources (Latané and Darley, 1968), and to be slow to connect a general emergency situation with a need to depart from their own personal status quo (Mileti, 1975; Kunreuther, 1979; Dynes et al., 1979). Frequently, people have become accustomed to false alarms, which leads them to discount the warnings they receive (Breznitz, 1976). Obviously, then, the effectiveness of initial responses to an emergency depends on the credibility of the source of notification. Since warnings from many private groups will be questioned by others, the public sector or the media must be relied on to supply objective and credible reports of what is going on.[3]

There is a dilemma: the more trust in government is needed, the less likely it is to exist (Axelrod, 1981). Trust is most important when people are least prepared for an emergency, but people's confidence in government may well be shaken if they are not adequately prepared. In addition, the situation will probably be further complicated by what appear to be conflicting messages from government (Axelrod, 1981). During a cutoff of oil imports, for example, just when government is trying to convince domestic parties that certain exceptional actions are needed, it can be expected to be trying to convince foreign parties that the United States is doing all right and that foreign pressure tactics are futile. At the same time that the U.S. public is being encouraged to adapt to the emergency, the government will be assuring people that there is no need for alarm; it may also be drawing on the strategic petroleum reserve, which would actually reduce pressure that leads to adaptation. At the same time that it is advising the country to rely on economic markets, it will itself be relying on trust in government. The government acts in this way when it assures the public that energy suppliers are not exploiting the crisis situation for unfair gain. It is worth remembering that during the 1973–1974 oil embargo, one of the first and most pressing requirements on the federal government was to investigate rumors that energy companies were hoarding oil (U.S. Department of Transportation, 1975.)

As a consequence, an emergency information system, if it is to be effective, must be responsive to public concerns as well as to government interests. It must allow the skeptical to resolve their suspicions, not simply provide facts: relying on the testimony of "experts" is clearly insufficient (e.g., Kash et al., 1976; Hoos, 1978). The system must be prepared to receive communications, not just disseminate them. Receiving communications identifies where credibility is a problem, points to unforeseen informational needs, and helps officials to refine their own perceptions about what is actually happening in response to an emergency.

Improving Emergency Communications

Given such a formidable challenge, it makes sense not to attempt to gather every bit of detailed information, but instead to ensure that communication lines are kept open. In a crisis situation, unusual kinds of information may be needed quickly. For example, there may be a need to know what legal authority state and local officials have, which vacant buildings are available for shelter, or whether school buses can be used to supplement mass transit systems. The requirements for such information—and the facts themselves—will usually be highly localized. It would be futile to try to maintain a flow of timely and correct information to meet so many localized needs. More appropriate is a system that focuses centralized attention on a limited number of information functions and that assures their credibility. It is important for the system to be an effective emergency *communication* system as distinct from an *information* system.

Meeting information needs in an emergency through a decentralized, pluralistic structure increases the importance of communication among individuals and among local organizations. Consequently, attention must be given to such prosaic matters as maintaining up-to-date lists of correct names and telephone numbers of key people. It is usually easy to identify ahead of time who will have a particular type of information, even if it is difficult to gather the information and keep it up to date. Also, because some communication links may be congested or broken in an emergency, the need for localized information suggests a need to build some redundancy into the emergency communication system to be used.

IMPLICATIONS FOR THE FEDERAL ROLE

The Dilemma of Federal Involvement

The more severe a future energy emergency is, the more certainly will the federal government play a major role in meeting emergency needs. This fact creates a real dilemma for the allocation of emergency responsibilities among levels of government and the rest of society. To provide any incentives for others to share in the process and burdens of prevention and preparedness, the federal government must convince the public that it will not do everything. But if other parties are to prepare, they do so in the knowledge that in an extreme situation, government may act to distribute the benefits of their preparations over the society as a whole. A policy of no federal preparations would force preparations and responses to be decentralized, but it would become politically untenable in a serious emergency.

There is also a significant danger that, in trying to shift some of the preparedness burden to the rest of society, the federal government may bring about the worst of both worlds: it may limit its own preparations as part of an effort to underscore the need for action elsewhere, yet fail to be convincing enough in this effort to actually trigger such action. If this happened, the federal role might be smaller than it should be in a severe emergency because of inadequate federal preparation; in a moderate emergency, it might be larger than it should be because of the inadequate preparation made by others.

A central empirical fact helps define the federal role: private parties seldom prepare adequately for emergencies (Kunreuther et al., 1978; Slovic et al., 1977). As a result, if private initiatives alone are relied on, many people and organizations will be unprepared (Kunreuther et al., 1978; Slovic et al., 1977, 1978; Meade, 1970).[4] When the federal government makes emergency preparations to fill a gap, those programs may deter decentralized preparations, and, as we have already noted, a decentralized approach seems necessary.[5] In addition, when the federal government can be expected to deal with the impact of an emergency, such as by offering disaster relief, others are unlikely to invest in disaster insurance or other individual preparations.

Experience demonstrates that when people are not prepared and suffer greatly, the government will tend to meet their needs even in the absence of stated policies; in effect, people who fail to prepare are rewarded as "free riders," while people who have prepared are penalized. It can be hard to persuade people that this experience will not repeat itself. And the more that people believe the government will intervene, the more necessary that intervention will become.

The "No Policy" Approach

In principle, one way for the federal government to maximize decentralized responsibility for handling an energy supply disruption is to offer no policy at all. If the federal government refuses to rescue individuals, private organizations, and local government, these groups will be forced to make their own preparations and find their own ways to respond.

The "no policy" policy does, indeed, seem an effective way to allocate scarce energy supplies. If energy supply decreases, with demand substantially unchanged, energy supplies will be quickly reallocated, at a higher price. Realistically, a very abrupt supply change would produce a period of energy shortages in an economist's sense, that is, situations in which supplies are not available for purchase at prevailing prices. Inflexibilities in routine pricing and inventory control systems would tend to have this effect, as would the strategic wariness of large energy corporations fearful that rapid price increases would trigger imposition of price controls. In some markets, the existence of long-term relationships between buyers and sellers might bring equity norms into play and result in attempts to hold prices down and allocate supplies with informal rationing systems. Of course, some prices are regulated, and their increase would be delayed and constrained by the regulatory mechanism. But, when these factors do not operate prices will rise rapidly, and it is likely that, given a sharp and obvious reduction in supply, prices will soon increase in most markets.

With sufficiently large price increases limiting the number of potential buyers, shortages would disappear relatively quickly, probably in a matter of weeks: that is, buyers would be able to buy as much as they wanted and could afford at the higher prevailing prices. This conclusion holds regardless of the extent of the price increases and thus regardless of the damage such increases would inflict on poor people and on society as a whole. Whatever sacrifice must be imposed to force energy quantities demanded down to the reduced levels of supply, price increases can impose it. If the market is left to operate, price increases *will* impose it. In this sense, allowing the market to operate may be regarded as a reliable policy for the full range of energy emergency contingencies.

Political Feasibility.   There are two broad questions to be raised concerning the general policy concept of allowing market forces to determine the nation's adaptation to energy supply disruptions. The first is whether such a policy is feasible. Feasibility is not, in this case, a question of the capabilities of government agencies; in that dimension, the demands of noninterventionist policy are small. The question, rather, is one of political acceptability. Undoubtedly, the answer to this question depends heavily on the severity of the crisis. Although the historical evidence from the 1970s indicates that emergencies of even such a moderate scale are sufficient to provoke substantial interventions, it is arguable that during those crises there was a political choice available. Perhaps reliance on the market is more clearly available as a policy option today, should a comparable moderate emergency occur in the near future. However, for more severe crises, there are strong reasons to believe that the pressures for political allocation of the necessary sacrifice would prove irresistible.

The defect of the market mechanism from a socioeconomic point of view is its indifference to equity. It treats energy simply as a commodity and ignores its status as a social necessity (see Chapter 2). While notions of equity are often ambiguous enough to have little impact, a national emergency tends to sharpen their meaning and create many specific equity demands. If there is a national emergency, should it confer great good fortune on some, while the remainder bear not only the full burden of the national misfortune but also the burden of wealth transfers to others? Would the public accept that some fiscally poor communities would be unable to afford enough fuel for health care facilities and fire and police protection?[6] It is noteworthy that regardless of the actual success of interventionist policies in mitigating inequities, the political stance of an administration that is visibly trying to achieve equity is vastly different from that of an administration that says it is *not* trying to do so. The force of these considerations in determining the actual policy course in an energy emergency should not be underestimated. Thus, even if a political decision is made to use a market-oriented approach to emergency response, the policy may fail the test of feasibility unless it is coupled with other policies and actions to ensure that critical social needs are met. If such needs are met, the prospects are much better that government will be able to limit its role and that markets will be allowed to operate.

Desirability.   A second broad question concerns the desirability of reliance on the market mechanism to cope with an energy emergency. The issues may be addressed under the headings of equity, efficiency, and compatibility with other national objectives: the concern here is with what the government should try to do on ethical normative grounds, rather than what political pressure may make it do. This is not the place for detailed consideration of the normative aspects of equity, but such concerns are legitimate: energy is, in part, a social necessity, and we believe that the equity aspects of an unfettered market solution to an energy supply cutback can be substantially improved upon by government policy. Very real difficulties arise, however, in defining and effectively implementing policies that are responsive to the vast diversity of individual circumstances. And it should be noted that well-vocalized and politically effective appeals based on equity and urgency of need may produce results that differ markedly from the relief that would be granted by an ethically sensitive and well-informed judge. In trying to design truly equitable policy, the political uses of equity claims are part of the problem.

Although efficiency is a virtue often claimed for market-determined resource allocation, the circumstances of an energy emergency may be such as to leave an important role for government policy in bringing about efficient adjustment. In part, of course, this is because the market is far from unfettered under normal conditions; rather it is fettered both by public policy and by private long-term arrangements that look advantageous in normal times but may not be so in an emergency. Many of the price-determining arrangements of the economy have considerable inflexibility. They do not respond nearly so well to new information as do, for example, organized securities and commodities exchanges. The economy simply does not maintain, under normal conditions, a set of market institutions that is well adapted to the peculiar short-term needs of an emergency. Market arrangements that work well under normal conditions may be quite inadequate to an emergency, as participants may need to find new and unusual transaction patterns or information sources. Self-interest may be expected to bring about needed adaptations eventually, but the market may function poorly for some time. Since these conditions prevail in different markets in differing degrees, and since prices are interdependent, it is unlikely that prices would move quickly to a configuration appropriate to guide efficient response to the emergency. Serious instability is not to be excluded, and a protracted period of "hunting" for a new equilibrium is likely.

The issue of compatibility with other national objectives recalls the fact that energy is not only a commodity but an ecological resource, a social necessity, and a set of strategic materials. The problem is most starkly illustrated by the possibility of an energy emergency combined with a military mobilization effort. The heart of the problem is the fact that the market is plainly not an appropriate institution to weigh the urgency of the mobilization or to determine its form. There are major complicating factors: the importance of government procurement, the fact that governments usually are not particularly nimble market participants, and the fact that many of the markets in which the government would seek to buy are far from competitive. There are further complications involving the macroeconomic problem of financing the mobilization. Taken together, these considerations strongly suggest that, in a future mobilization as in past ones, the government would have to intervene strongly in the economy to bring about an economic response compatible with national objectives. In such a context, much of the efficiency that markets display is likely to be efficiency in support of objectives contrary to those selected by the government.

The compatibility issue applies to civilian priorities as well. The political process often makes decisions based on judgments of energy needs or national economic security. Thus, if it is agreed that there is more need to heat houses of the poor than to run the snow-making machines at ski resorts, direct policy intervention in the market might be needed. Similarly, if the nation decides it needs all the primary aluminum it can get, it would be appropriate to override a market signal that was shutting the industry down.

It is to be emphasized that the force of the objections to reliance on the market increases with the severity of the emergency. While government passivity might effectively encourage adaptive decentralized preparations for an emergency of moderate severity, that strategy would be less tenable and less appropriate for a severe emergency.[7]

It has been noted that the federal government's functions in an energy emergency include seeing that social needs are met and that national security is maintained. There is also a federal role in linking national, regional, and local emergency preparedness activities (Kreps, 1978). In most energy emergencies, one can expect states and localities to act to take care of their own interests or problems. In an oil import crisis, for instance, some oil-producing states may seek to set aside a substantial part of their production to meet state needs before providing oil for the general market or other states. Other states may try to enact restrictions on energy use, even if the federal government does not. Some of these actions are a natural part of a decentralized system for response, but some of them could increase regional differences in the impact of an emergency and, as a result, lead to conflicts between regions. Uncertainty about government actions to resolve interstate and interregional conflict can undermine attempts at the federal level to clarify public-sector intentions. Certainly, state actions could present the federal government with a host of legal issues and challenges, many of which might best be resolved before an emergency occurs. Current energy emergency planning at the state level may provide clues to disputes that could arise.

IMPLICATIONS AND RECOMMENDATIONS

Decades of research in the behavioral and social sciences have produced a number of findings that can be used to plan for reducing the threat of energy emergencies. On the basis of insights gained from these studies, several principles should be recognized as part of a sound U.S. policy toward energy emergency preparedness, and several actions can be taken in accordance with those principles. Although these recommendations cover only a part of the territory discussed in this chapter, they can be a significant step in the right direction. The points are organized under three headings: federal government roles, organizing for preparedness, and information functions.

Federal Government Roles

Wherever possible, the federal government should take full advantage of the flexibility inherent in decentralized emergency preparation and response. The federal government should allow and even encourage individuals, groups, and areas to seek and use alternative response strategies. Rather than focusing an incentive program on private oil stockpiling alone, the same incentives might be made available for other investments that offer the same benefits to the nation. This approach would invite parties to be creative in identifying options that they believe make sense for them and for the total energy emergency program. Decentralized efforts to improve energy efficiency during nonemergency times constitute one important option because such efforts can help prevent emergencies. Besides encouraging innovation in the U.S. society and economy, promotion of decentralized responses can help build a more diversified and resilient system for responding to emergencies. As a first step, the federal government should ascertain the potential roles of social groups and relationships in a decentralized approach to emergency preparedness and help to identify options for enhancing their potential.

To the extent possible, the federal government should ensure that emergency preparedness is clearly linked with emergency prevention.   Not only can preparedness help to prevent some kinds of crises, but a full portfolio of preventive actions can help to reduce the challenge of emergency preparedness—which is formidable at best—and to make preparedness more credible as a public and private enterprise.

In its troubleshooting role, as part of the endeavor to rely on decentralized responses as much as possible, the federal government should ensure that critical social needs are met during energy emergencies.   If critical social needs such as health care and law enforcement are not met during an emergency and if major inequities result from scarce energy supplies and higher prices, there will be strong, perhaps irresistible pressures for centralized government action. Even an administration that wishes to sustain a market-oriented, decentralized approach to an emergency must rely on previous federal government action to ensure that critical needs are met and that inequities do not reach a socially unacceptable level. As a step toward these objectives, a careful study of the needs of groups affected, of essential services required, of conditions under which services would be affected, and so forth is necessary, and alternatives for meeting these needs should be identified and examined.

The federal government should keep to a minimum any uncertainty about its potential uses of the policy instruments available to it.   Whenever there is uncertainty about the potential uses of an emergency power by the federal government, such as price controls, fuel allocation policies, or decisions to sell oil from the strategic petroleum reserve, other parties tend to assume that full government power will be used. As a result, many of them will take less responsibility for emergency preparedness themselves, increasing the demands on government during an emergency. In this sense, uncertainty about government policies has a strong policy impact; clearly, if this impact is not what the government would wish, the uncertainty needs to be reduced. The federal government should work toward broad nonpartisan agreement on the roles and functions it will perform during an emergency, so that uncertainty and skepticism are reduced. For example, in the case of an oil import cutoff, agreement on such matters as the mechanisms to be used for a drawdown of the strategic petroleum reserve and for revenue recycling is vital. Since such actions require resolute cooperation by the various branches of the federal government, they present a substantial challenge.

The federal government should act to reduce uncertainties about federal– state relations during a major energy emergency.   Through legislation or court action, many of the current uncertainties about state powers in an emergency could be reduced substantially, and some legal disputes during an emergency could be avoided. At this stage, the important thing for emergency planning is that the issues be resolved, not that the resolution take a particular form. As a first step, the federal government should identify conflicts that might be caused by state and local emergency actions, including possible conflicts between energy-producing states and energy-consuming states. The federal government should seek to resolve these important issues before there is an emergency.

Organizing for Preparedness

Preparedness should be based on a continuing process rather than an infrequent production of "plans."   Because contingency plans, once prepared, are seldom used, it is more effective and more efficient to create a continuing process than a set of documents. For instance, recurring tests and exercises can be used to evaluate the nation's preparedness; to train people in key positions; to train people in the use of communications systems; to test the readiness of these systems, for example, to check on whether lists and telephone numbers of key actors and information sources are up to date; and to investigate the nation's options with respect to different contingencies. In addition, the federal government should seek to learn more about the importance of such factors as leadership, group norms, and media attention in maintaining public support for emergency preparedness.

Preparedness should focus on systems for response rather than the responses themselves.   Because it is so difficult to foresee the full range of characteristics of a real emergency, a system capable of responding to unexpected circumstances is better than a system that consists mainly of prespecified actions. The system has to be able to learn from people's and institutions' behavior during the early days of an energy emergency. If the system is locked into a pattern of response that was defined well in advance, the policies may be poorly suited to the realities of the situation. The federal government should develop an organizational structure for energy emergency preparedness that is broad-based, related to a wide variety of possible contingencies, and linked with complementary structures at the regional and local levels.

The nation's energy emergency preparedness effort should not be limited only to oil import interruptions.   It would be a mistake to set up a system for emergency planning and response that is tailored exclusively for an oil import interruption. An oil import emergency might lead to other energy emergencies, but other emergencies might arise first. Preparedness should be broad-based, capable of handling such events as a severe electricity shortage, a major strike, a natural disaster, or a major military mobilization, as well as an oil import cutoff.

Energy emergency preparedness programs should be concerned with post-emergency recovery.   This recovery, as well as crisis management and impact mitigation, must be part of any plan because recovering effectively from one emergency may be essential in avoiding or preparing for another.

Information Functions

Federal programs to meet public information needs in energy emergencies should directly address credibility issues.   The efficacy of both centralized and decentralized decision making during an emergency is likely to depend on the credibility of available information about the emergency. Unless information can be verified and thus widely accepted, it may tend to increase uncertainty and divisiveness rather than to reduce it. The federal government should investigate how to design a public information program that is credible to the public. Such a program would need to include ways to respond to public concerns and, in some cases, mechanisms for verifying the information provided.

Aside from providing general information about what is happening, the key federal role in an energy emergency is ensuring that communication systems are ready to meet emergency needs.   There is no way that the federal government can meet all the needs for detailed information that will arise in an emergency. Most of those needs will have to be met in a decentralized way. But the requirements for coordination that are inevitable with decentralized action will increase the importance of effective communication systems, capable of operating under emergency conditions. The starting point is such basic matters as lists of key personnel, both public and private, and how they can be contacted in a hurry. But the challenge also includes providing alternate means of communication in case the regular system is congested or interrupted. Whether or not the federal government must itself provide such systems can be debated, but one of its most critical functions is to ensure that such systems exist.

The federal government should improve its knowledge of the information needed during an energy emergency. It should determine what types of information are needed, the best sources for providing these types of information under emergency conditions, and the likelihood that those needs will be met without federal government initiatives. The federal government should consider what kinds of information can be stockpiled ahead of time for an emergency. For instance, most kinds of detailed information will quickly become obsolete, but prepackaged news releases and television spots that provide examples of coping strategies, such as options for families if electricity supply is interrupted or petroleum products are very scarce, could be very useful.

By adopting such principles, policies, and actions, the government of the United States could make the prospect of energy emergencies less threatening to the social and economic structure of the country. In addition, the process of improving readiness in this way can help the nation redefine the appropriate roles of central government and decentralized decision making.

Notes

1. Of course, certain problems must be resolved in a centralized way. For an oil import emergency, examples include the use of the strategic petroleum reserve, dealing with restraints on interstate commerce, and meeting energy requirements for national defense. And local options and needs may be strongly affected by prior actions by central authorities (see "Implications for the Federal Role," page 155, and "Prevention and Preparedness," page 138).

2. This attention to information can mean that it is collected differently, introducing bizarre noncomparabilities into data series.

3. At the time of initial notification, certain parties who believe they have superior information may define the situation as an emergency while others without that information, or questioning the validity of the information, do not. Seeing the same response from the public, one group would perceive underreaction while another perceives overreaction (see "Diversity and Conflict," page 141).

4. This unpreparedness does not occur because private parties are unintelligent or fail to foresee events as well as government. But while a private party may carefully consider benefits and costs to itself, it does not usually take full account of the benefits of its actions for others. Consequently, private initiatives may pay too little attention to actions that make sense for a larger group.

5. Some government programs do not necessarily deter decentralized preparation, for example: preventing emergencies by encouraging shifts to alternative energy sources and supplying information about national fuel supplies. Preparedness activities are particularly likely to deter decentralized actions, and stockpiled capabilities can have a similar effect.

6. A severe energy emergency poses such questions even more sharply than mobilization for war, because most of the beneficiaries of the crisis form a narrowly defined group whose members are readily identified—as contrasted with the more diffuse group of wartime "profiteers."

7. This is not to imply that in a severe emergency, any action the government might take would necessarily be preferable to government inaction. Because of the diversity of emergencies, we have not attempted here to judge the desirability of alternative government policies for severe energy emergencies.

7

Local Energy Action

Over the past decade, more than 2,000 U.S. communities have taken actions to produce or manage energy and energy services to meet local needs (see, e.g., Center for Renewable Resources, 1980). Our committee was drawn to examine these local energy actions because of our interest in social processes and institutions between the levels of individual decisions and of federal policies and programs. On the basis of our understanding of these processes and institutions, our working assumption was that significant opportunities might exist under present technological and economic conditions for local communities to solve their own energy problems. We also expected that even when the technical and economic environments were favorable, local energy activity might be beset by many serious problems.

Our analyses of other energy issues, presented in preceding chapters, increased our interest. We came to see that local institutions are often in a better position than either the federal government or the market to cope with certain important energy problems: for example, local institutions may be best suited to take the lead in the transition to more efficient energy use (see Chapter 4). Unlike the federal government and energy-related industries, which often have credibility problems, local groups are in a good position to gain the attention and trust of local audiences. Compared with national institutions, local organizations are in close touch with local needs and concerns, are potentially controllable by those who need their assistance, and are in a better position to assess the quality of service available from local energy businesses, such as heating or insulation contractors. At least in principle, local groups could be the most effective institutions for increasing energy efficiency in the interests of energy users and the nation. Local institutions are also in a good position to transmit information to and receive information from the informal social networks that are so influential with energy users.

Local institutions may also be best suited for doing some of the necessary work of planning for and responding to energy emergencies (see Chapter 6). Because emergencies affect different geographic areas in different ways, local authorities will be in the best position to identify and respond to local needs. As we have argued, local institutions should be centrally involved in emergency planning, for unless they are involved in the planning, they will not know how to respond effectively. Furthermore, an understanding of local energy management during stable or normal times may have value for energy emergencies: to the extent that local energy systems can act independently of each other, the society as a whole is less vulnerable to any localized disruption.

There are many good reasons for thinking that more local control could, at least in theory, improve energy management both under ordinary conditions and in emergencies. The critical question is whether the potential advantages of local action can be realized in practice. Although local energy management may require different scales of technology, it rarely requires new technologies, so feasibility is mainly a social, political, and organizational question. Unfortunately, local energy management has not been carefully studied by energy analysts, possibly because local management relies heavily on institutions other than the federal government and the market.

This chapter examines local energy action as a set of phenomena. We note the existence of sharply differing views of its feasibility under present conditions and the lack of sufficient knowledge to make a reliable judgment. We then offer a partial framework for learning more about the potential for local solutions to energy problems.

THE PHENOMENA

Since the oil embargo of 1973, an increasing number of local governments and organizations have struggled on a path that they hoped would lead to reducing the demand for fuels and electricity, to expanding the use of alternative methods of producing and delivering energy, and to mitigating the effects of fuel price increases in their areas. We refer to this collection of phenomena as "local energy action;" we define it as collective action at the local level to meet local needs for energy services. Energy services— that is, heating and cooling, mobility, industrial processes, and so forth— can be provided by any combination of conventional fuels, renewable energy resources, and energy-efficient technology.

Local energy actions vary widely—in sponsorship, in content, and in the ways they have come about. A few examples indicate their range. In

1979, representatives from the federal agency, ACTION, approached leaders of the city of Fitchburg, Massachusetts (population, 39,000) with a proposal to conduct a program of low-cost weatherization of residences in the city. The result was a six-week crash program that used $42,000 in money from two federal agencies and the state energy office, substantial donations of volunteer labor by community members, and in-kind donations of space and materials. About 60 percent of the residents of the city took some energy-conserving action during this period, saving an estimated 14 percent of their energy bills (Fitchburg Office of the Planning Coordinator, 1980; Stanton, 1982). With this evidence of success, ACTION and the Department of Energy began to spread the Fitchburg model around the country. In Fitchburg, an independent agency was created to continue the work, and the city became interested in applying the techniques it had learned to its public buildings.[1]

In the San Luis Valley of Colorado, low-income Hispanics who have lived there for generations created People's Alternative Energy Services. The organization works to help people reduce energy costs while building local self-reliance. It runs workshops to teach passive solar techniques to homeowners and produces do-it-yourself texts in English and Spanish on such subjects as passive solar applications for adobe buildings. When it builds passive solar greenhouse additions to houses, the homeowners are required to help with the construction. The group emphasizes self-help, but it has benefited from grants and gifts from outside groups (Stern, Black, and Elworth, 1981).

Davis, California is famous among localities around the country for its pioneering energy activities. One of these, enacted after lively debate, is a building code that virtually requires passive solar design in new residences (Brunner, 1980). Builders objected at first, but soon found that with very little extra cost for construction they could build homes that meet cooling and heating needs at considerably lower cost for energy than their standard homes. And the effects of the building code went beyond the energy saved in new buildings. In the first forty-one months after the building code went into effect, electricity demand throughout the city dropped by 15 percent (Dietz and Vine, 1982).

In Auburn, New York, the mayor has pursued an aggressive production-oriented energy policy. He has developed and promoted local natural gas deposits and initiated work on local hydroelectric and geothermal projects. The goal of the small city is economic development in an economically declining area of the northeast. Local industries, schools, and hospitals have taken up the search for natural gas and as a result have done much to cut their energy costs (Cose, 1984).

The list of examples could go on at great length and with great variety. Energy activities have been undertaken by local governments and by grass-roots groups. They have been oriented toward conservation, renewable energy, and traditional energy production. They have been designed to benefit homeowners, renters, commercial business, farmers, motorists, and municipal budgets and to improve the general economic climate of an area. Sometimes they have been locally initiated, but sometimes the impetus has come from the outside. Some projects involve expensive technologies, others rely mainly on volunteer labor, and still others are regulatory in nature. Some projects show dramatic results, while others never proceed far enough to produce any net energy benefit. What almost all these efforts have in common, unfortunately, is that they have not been evaluated to measure their effects.

HOW MUCH CAN LOCAL ENERGY ACTION ACCOMPLISH?

Although thousands of communities have taken action to improve their energy situations, local action has never been a major concern in national energy policy debates. Congress has often expressed support for local energy action, but it has provided much more funding for synthetic fuels development and other national-level efforts than for resources to local groups for their energy activities. And while the Reagan administration's recent statements on domestic policies have included many positive references to the ability of local institutions to solve problems, the administration has rejected efforts to provide federal resources for local energy action—it took a federal court decision to force release of the resources legally mandated in the Conservation and Solar Energy Bank.

Ambivalence also exists in our committee. While there is general sympathy with the idea of popular control of institutions, there is sharp disagreement about whether more local control of energy would actually produce more efficient or flexible energy systems, and there is also disagreement about whether it would be more "popular" than the present system.[2] Since the debate cannot be resolved by facts, we present the arguments in capsule form to highlight the points of disagreement.

The Case for Optimism About Local Action

Local energy action has the potential to reshape the national system of energy supply and use, while simultaneously helping to solve some other national problems. Because many social trends are in the same direction—away from large and distant institutions to solve local problems—there are strong reasons to believe that this potential can eventually be achieved. For example, local energy action makes sense as a way to use regional and local energy resources more efficiently. It can make fuller use of local information and capabilities, including unrealized potentials for increased energy efficiency. A number of recently developed energy technologies, especially in the areas of renewable energy and conservation, are particularly suited to local application.

Local energy action offers a way to increase local self-reliance and control at a time when American society is questioning the performance and motives of large institutions, both governmental and commercial. Some pressing energy problems may be solved through local self-help, and, at the same time, the effort helps build a community's ability for problem solving. At the very least, the experience will train a local group of experts in energy planning and management that can help a community make quick and appropriate decisions in an emergency situation. Activity on a local energy project may also increase the general sense of solidarity, cooperation, and trust in a community. This is a frequent outcome when social groups are interdependent on a common set of resources (Sherif, Harvey, White, Hood, and Sherif, 1961; Stern and Kirkpatrick, 1977), and was reported by the Fitchburg Office of the Planning Coordinator (1980).

Local action may be the best route to solving some of the problems energy prices have created in many communities: the abandonment of housing stock, the declining competitive position of local business, and increasing costs for municipal services, to name only a few. Energy problems affect different communities in different ways, and neither the federal government nor the market is well organized to meet all the particular needs of individual communities. At the local level, these needs can be most accurately identified, and there can be public debate over ways to meet them. Local energy conservation and production activities are among the possible solutions for the problems the national energy situation has created at the local level.

Local energy action can also be a way to produce much-needed innovation in the national energy system. Communities may be motivated by special local conditions to try programs or policies that would seem inappropriate as national policies. Communities can thereby benefit the nation by developing a body of experience that can be shared with other localities when useful and that can be applied to emerging conditions on relatively short notice. A greater number of potentially useful strategies are likely to appear than if the nation relies only on national decisions and policies. The spontaneous adoption in past oil shortages of "odd-even" gasoline rationing—a method of rationing gasoline based on the last numeral of the licence plate number, which began spontaneously in a few states—is an instance of the adoption in many parts of the country of one successful solution.

Local energy action is a way to make the national energy system more stable and resilient in the face of rapid changes in world energy conditions. We noted in Chapter 6 that diversity in the national energy supply and distribution system makes it less vulnerable to disruptions in the supply of particular fuels and therefore is a form of emergency prevention. By promoting diversity, local energy activities can contribute to solving national problems of vulnerability.

Finally, local solutions to energy problems may be easier to achieve than national solutions. At a time when the United States seems unable to sustain decisions about energy policy at the national level, local action offers a chance to weigh the trade-offs among energy, environmental, economic, and other objectives at the scale at which stable consensus or coalition formation is a realistic prospect. Local energy action can link the concerns of local businesses with economic growth and the interest of local community activists in public participation and the needs of the disadvantaged. It can serve as a catalyst for new partnerships at the local level between the public and private sectors, rather than sparking conflict between them.

The full potential of local energy action has not yet been reached. This lack is partly due to the experimental nature of the current situation: local energy actions are a new response to new conditions. If the lessons from recent experience with these activities can be spread, the rate of success might dramatically increase in the future.

There are problems associated with local energy activities, to be sure. They can lead to conflicts between neighboring localities and they will fail to meet certain needs of society as a whole (Wilbanks, 1983). Local actions that succeed in meeting energy objectives, such as supplying or saving a certain amount of energy, may be less successful in meeting social objectives, such as public participation. And local action will seldom occur in complete independence of outside sources of support. To focus on the limitations of local energy action, however, is to miss the main point: that the potential of local action is indeed substantial, large enough to have a truly significant impact on the national energy future. The real question is not *whether*, but *how* to achieve the potential.

The Case for Pessimism About Local Action

There is no question that if local governments and institutions throughout the country could take effective action on their energy problems, many of the beneficial results claimed could be achieved. But there is little reason to believe that under present economic and political conditions, widespread local success in energy activities is likely. Neither is there much reason to believe that local action would often improve the situation of poor people, as is sometimes claimed, or that consensus would be easier to achieve for local energy policy than it is for national energy policy.

Claims of past success for local energy activities have probably been exaggerated. Many programs that are glowingly reported in the planning stages are never heard from again; they are likely to be the failures. In addition, programs that fail are less likely to acquire the resources to evaluate themselves than programs that succeed, so most evaluations will be of successes. Furthermore, most reports of the results of programs are prepared by someone with a stake in making the program look good: such reports are probably too quick to interpret change as success and too quick to attribute success to the program rather than to other events. Because careful and impartial evaluation studies of local energy activities are done so infrequently, it is rarely possible to rule out the interpretation that reported successes are simply local manifestations of a slow national trend toward improved energy efficiency.

It is also questionable whether activities that succeed under a particular set of local conditions can be easily adapted in a different local area. More than local energy conditions are involved in making an energy program work: even when the energy needs of two communities are almost identical, the political realities may differ so much that what works in one community cannot even get serious consideration in another.

To make matters worse, very few local energy activities have gone far without federal financial support, and such support is not now forthcoming. The same can be said for indirect federal support in the form of information, research on problems of concern to communities, and support for the individuals and groups that provide communities with ideas and expertise. There is little reason to believe that the few documented successes among thousands of local energy activities are the first signals of a trend with significant national potential.

Examination of the politics of energy gives no reason to believe that what national politics has not been able to accomplish would be easier to accomplish at the local level. The political alignments and difficulties are essentially similar at both levels. For example, the interests of energy producers and energy consumers operate at both levels, and are usually antagonistic. At both local and national levels, there have been attempts to use energy policy for redistribution of economic resources, and at both levels these efforts have met powerful resistance. Nationally, the major conservation programs that have been implemented—tax credits and the Residential Conservation Service—benefit primarily the wealthier consumers, and the limited evidence that is available suggests that local programs have been no more successful than national ones in achieving goals related to social and economic equity. Under present economic conditions, energy needs cannot be met by economic expansion—and this problem is at least as severe at the local level as it is nationally. Declining levels of federal support for local services only intensify the competition for the remaining local resources, and under those conditions the dominant forces in local politics are unlikely to cede much to newly organized interests. In short, the political problems that have proved so difficult in national energy policy also exist at the local level, and the forces in opposition are just as strong. Great progress in energy policy at the local level is unlikely until the national political-economic picture changes.

Resolving the Argument

The striking fact in the argument about local action is that, despite reports of thousands of local energy activities in the United States over the past several years, we cannot be sure which of the above cases is closer to the truth. The research to date has been insufficient to make such a judgment. The experience and capability exist to learn much, but the will to do the research has been lacking.

A discipline of evaluation research that can provide fairly rigorous scientific analysis of social innovations, such as local energy activities, has developed over the past decade. While this discipline has been used to evaluate a variety of social programs, it has only infrequently been used to examine energy activities. It is hardly ever used to examine energy activities that are neither funded nor mandated by government.

This inattention to local energy action may be due simply to the recency of the phenomenon, but it may also reflect, in part, the way many analysts think about energy. The usual concerns of energy analysis are technological research, development, and demonstration; evaluation of economic feasibility; and analysis of possible federal policies. Since local action rarely results in new technology and hardly ever proceeds unless the economic feasibility is fairly clear, local action may be uninteresting to many policy analysts. Worse, most energy analysts may implicitly assume that all the nation's energy opportunities—and its problems—are technical or economic; that once technology is developed and economic analysis is complete, local decision makers will fairly predictably choose to implement whatever is efficient in their contexts.

We do not agree with this view. We see local energy action as a difficult social and political undertaking, even when the technical and economic issues are clarified. But we are unable to draw conclusions about the potential for local action because of the scarcity of careful research on outcomes and the almost total absence of studies of the processes by which local groups address their energy problems and by which their efforts proceed, stall, or become transformed. The lack of careful research and evaluation is unfortunate because interest in the topic among the public and among policy makers at local and national levels seems to remain high. Careful research on local energy activity will be the best way for the public and officials, locally and nationally, to learn what potential exists.

The next section identifies some issues we believe are likely to be important in the experience of local energy activities, issues that are worthy of careful study. Our judgment is based on existing knowledge about social and political processes at the local level and on the limited knowledge of local energy actions available from the literature and contacts with practitioners in the field.

ISSUES AFFECTING LOCAL ENERGY ACTION

Getting on the Agenda

Traditionally, energy is not a concern of local governments or community organizations. Local energy action does not occur unless energy first captures the attention of some organization capable of initiating activity. For this to happen, energy must first be defined as a public issue. One barrier to that definition and attention is the view of energy as a commodity (see Chapter 2). Most people's experiences with energy are closely tied to their experience as customers of private energy companies. When they have problems related to energy, they are likely to look for redress to the energy producers or to the state and federal institutions that can regulate them, rather than to local government or community groups. Since governments and community groups rarely produce commodities, the commodity view reinforces the belief that local actors have no role to play in developing energy policy. This usually leaves energy in the hands of the private sector institutions that are already making most local energy decisions.

Energy activities may get more attention in communities where energy is already a public issue, such as cities with municipally owned utilities. In these communities, however, the shape of past energy actions may determine what occurs next. For example, in the mid–1970s, technical analysts for the municipally owned company, Seattle City Light, projected increases in demand for electricity that would require additional generating capacity by 1990. The utility recommended solving the problem by the usual routine—investing in new power plants. The recommendation would probably have been adopted, but a citizens' advisory group questioned the demand projections and called in outside consultants to develop another estimate. When these experts projected a lower level of demand, the advisory group recommended, and the public approved, a citywide conservation program that could save money by making the new generating capacity unnecessary (Brunner, 1980).

The Seattle example demonstrates that energy activity and inactivity both have their own momentum, but that such momentum can be reversed. Public control may increase the responsiveness of communities to new energy conditions because local political interests have the right to participate in energy decisions, but it does not guarantee that all issues will be debated or that all views will be represented. It seems likely that in the absence of strong proponents of the social necessity view of energy, or of a community political structure that includes strong representation of consumer interests, energy decisions will tend to be governed by the commodity view and by technical concerns and will usually be made by experts and their employers in government or business.

In recent years, communities without public energy agencies have frequently taken action on energy issues. In Los Angeles and some other communities, an energy shock with strong local effects triggered the action (Acton and Mowill, 1975). In many communities, energy issues got on local agendas because of a change of perception—it was seen as a way of meeting other needs, such as for employment, housing, or environmental protection. For example, weatherization programs have gained local support for various reasons only indirectly related to energy: advocates of low-income groups often see weatherization programs as a way to improve housing stock and provide job training and jobs for the unemployed. They also emphasize advantages for city government: weatherization may save welfare costs over the long run and prevent abandonment of rental housing. These are both benefits for local tax rolls. Homeowners of moderate income also find weatherization assistance attractive because cutting energy costs may ease their other economic problems. In communities suffering economic decline, production of energy from local resources has been attractive. In Susanville, California, when shutdown of a prison and an army depot created a local economic crisis, one response was a local geothermal energy project aimed at creating jobs and attracting business (*City Currents*, 1982).

This discussion suggests four propositions that are worthy of careful examination in light of recent experience: 1. Communities that are routinely involved in local energy action are more likely to undertake new activities than communities that have not accepted energy as an appropriate area of public activity. 2. Where there are local public agencies that address some energy issues, arguments for public action on other energy issues are more likely to be made and may receive more attention than in communities where energy decisions are completely in private hands. 3. One successful local energy action will make it easier to attract community attention to others, and so localities are most likely to get involved in energy by stages: an initial, minor involvement may facilitate further local action by a process analogous to behavioral momentum in individuals (see Chapter 3). 4. The ability of local groups to present energy action as a solution to nonenergy problems may have been significant in getting energy on local agendas where it had not previously been a public issue.

Mobilizing Action

The success of local energy action depends on mobilizing support by tying energy issues to philosophical views that grip and move people, by relating energy to concerns of organizations that already have some local following, and by building coalitions.[3]

When local groups propose new energy activities, at least five general philosophical outlooks have been presented in support. These "philosophies," which imply social objectives as well as energy goals, are important because they help define energy activity in terms of values legitimately pursued by local groups, and this gives such groups a reason to become involved. A "conservation" philosophy implies a definition of energy as a resource (see Chapter 2). Its proponents emphasize careful use of scarce resources and advocate moderating the demand for energy. A "community control" philosophy emphasizes decision-making power and how it is distributed. Proponents seek to democratize decision processes in general and decisions about the production and distribution of energy in particular. An "equity" philosophy focuses on energy as a social necessity (see Chapter 2) and stresses the redistribution of energy services to meet minimum needs. A "community development" philosophy is related in part to the previous two: some proponents emphasize revitalization of neighborhoods, using energy action as a way to meet social needs for employment, housing, and so forth; others put their major emphasis on the general benefits to the community of increased economic activity. Finally, an "appropriate technology" philosophy is most explicitly related to the antinuclear and limits-to-growth movements and what have been described as "soft energy path" models (Lovins, 1977). Its proponents support local approaches on the belief that they will lead to energy technologies of a smaller scale, which will be freer of the environmental and social problems of larger-scale energy technologies.

Philosophical support for local energy action becomes politically significant only when it is embodied in organizations. Using a community development philosophy, the chamber of commerce in Richmond, Indiana organized the people who provided essential support for the city's initial energy planning activity (Cose, 1984). This group built a broad base of support by emphasizing the idea that energy planning would benefit the entire city. In San Bernardino, California, the Westside Development Corporation began its energy activities by demanding funds from the city government—on equity grounds—to meet the needs of poor people. The funds were used to train people to produce and install solar energy systems in low-income housing units.[4]

The history of a local energy proposal will depend on the philosophical basis of the proposed activity and the distribution of its likely costs and benefits. The distinction between distributive and redistributive policies is useful in understanding this point (Lowi, 1964, 1972). Distributive policies have no clear losers, but some clear gainers: a benefit is offered, and potential recipients compete or negotiate to enlarge their shares. A good example is the Community Energy Project, initiated by ACTION. This project, modeled on the successful project in Fitchburg, Massachusetts, relies heavily on volunteer labor to run local weatherization efforts of short duration (Community Energy Project, 1981). Each community designs its own program; the programs run for low cost, and they offer widely distributed benefits. In fact, to get general assent to a project, it has sometimes been necessary to make it more distributive: to make sure that services are available to particular broad clienteles, such as homeowners ineligible for federal low-income weatherization assistance or occupants of small commercial buildings. Distributive policies tend to preserve the existing centers of power and they are therefore less likely to generate strongly organized opposition than redistributive policies.

Redistributive policies take from some to give to others. They tend to be more politicized because there are losers and because they usually form some identifiable interest group. Changes in utility rate structures are a clear example: a given level of revenue is reorganized so that what is a rate decrease to one group of customers is a rate increase to others.

Some of the impetus for local energy action comes from its redistributive potential. Advocates of the equity and community control philosophies favor energy activities because they can redistribute economic benefits or political power to less-advantaged groups in the community. But conflicts arise over redistributive policies when the potential losers anticipate the effects the policies will have on them and when they are sufficiently organized to act as political interest groups. Anticipation and organization are more likely to arise among energy producers and other business interests than among low-income consumers or other diffuse groups of people, because the former are often already organized to protect their economic interests.

This analysis suggests several propositions about local energy action that deserve study: 1. Energy actions tied to a community development philosophy and offering some tangible benefits to well-organized local constituencies are most likely to be given serious attention by local political bodies. 2. Energy activities with social-change objectives more often have to operate without the assistance of local government. 3. Proposals that would redistribute resources from powerful business interests to small consumers or to the public sector tend to give way to energy options that are less redistributive and therefore generate less organized opposition. Even after a plan of action is accepted, if a group with essentially redistributive goals decides that the action is not serving its constituents, it is likely to redirect its efforts to some nonenergy issue or use its support as a political bargaining chip rather than as a substantive commitment. In the process, the energy activity changes. 4. Redistributive policies are seriously considered only where the interests of consumers and lower-income people are well organized. Under these conditions, policy processes can be expected to become more participatory and also to create more conflict. 5. Economic stringency tends to narrow local energy options dramatically, because redistribution is politically difficult and distribution is limited by resources. Few feasible options may remain: improved management of public buildings and equipment, low-cost and no-cost building weatherization programs, and increased political conflict over redistributional proposals.

Obtaining Resources

Local energy actions typically require three kinds of resources: money, labor, and expertise. Clearly, different sorts of energy activity need different resources. Some activities, such as district heating or waste-to-energy conversion plants, are capital-intensive and depend heavily on an initial block of funds. Others, such as home weatherization, need a steady supply of labor. Expertise is important because most local energy actions involve technologies whose characteristics are not self-evident to most people, and because the policy options have uncertain costs and benefits. The mixture of resources needed probably also depends on whether the proposed option is new, seldom tried before anywhere; unfamiliar at the local level; or technologically sophisticated.

Few local energy action programs have been started without some outside resources. Substantial financial investment has sometimes been essential, but very modest inputs have also proved important—a small grant, interest from a federal agency, or the volunteered time of an outside expert. Such external resources can multiply their effect by attracting the attention and resources of local groups. A federal agency's offer of funds for weatherization to Fitchburg, Massachusetts attracted the attention and effort of the mayor and city planning department, and those central actors were able to attract local resources in the form of volunteer labor, donated building space, and the like (Fitchburg Office of the Planning Coordinator, 1980).

Although outside resources have been helpful, even essential at times, they can also bring problems. Considerable experience with technical assistance since World War II has taught that there are limits to the ability to transfer technology (see, e.g., Sutton, 1968). Local situations vary widely from what is thought to be a typical case, and it is difficult for an expert to get specific enough information to make useful suggestions for unique local circumstances. When a technical expert offers analysis, local groups may want action. When an expert explains how technologies or techniques work, local people may want to know how to use them in the specific locale. As a result, local people may resist expert advice—and their resistance may reflect an accurate appraisal of local realities. Furthermore, if a locality is interested in energy actions as a way to increase self-reliance, accepting technical assistance may be seen as an admission of inadequacy, which can lead to local resentment of outside experts and suspicion of their motives.

Another problem with outside resources is that they often flow to localities with strong organizational capacities to obtain grants and other assistance rather than to the localities in greatest need; this has been true of a range of federal programs in the past, from community development to water pollution abatement. Communities with strong organizational expertise and resources respond to available programs more frequently and are more successful in their applications (Friedman, 1977). Outside support also constrains what a locality can do, because local groups tend to choose actions that they believe can receive support, or because sources of support set guidelines for action.

Local groups sometimes achieve a great deal by drawing on their internal resources. The experience of Fitchburg is typical in this respect, and has been reported in many other localities that have received small grants (typically $5,000) as part of ACTION's Community Energy Project (1981). Other localities have also drawn heavily on internal resources. For example, the Pembroke Solar Project in Kankakee County, Illinois, has built a program to educate the local public about low-cost ways to meet energy and food needs, and has trained unemployed members of the community using only small grants for passive solar-heating technologies. As outside resources have become even more limited, the project struggles on with smaller self-help projects in an effort to become self-supporting.[5] In Richmond, Indiana, energy planning that was done with federal funds has led the city to choose energy options it expects to implement solely with internal sources of support, such as low-cost residential conservation activities, city-funded conservation education seminars, a volunteer program to teach businesses how to save energy, and economic research on the potential for district heating.[6] And the experience of success in Fitchburg has led the city government to apply weatherization techniques to its own buildings, and has encouraged a local community service agency, United Neighbors, to conduct workshops on installing solar collectors.[7]

Despite these hopeful signs, the main effect of decreased federal support for local energy action is likely to be negative. The strain on local budgets will increase and therefore make most new activities unrealistic. Existing programs can also be hurt. If local governments must choose between an existing weatherization program and traditional social services, weatherization may appear less affordable, even though it probably pays for itself many times over in decreased welfare payments, increased tax receipts, and so forth. Because social services are delivered continually and weatherization is not, recipients are more likely to notice a declining level of social services and to organize to prevent it. Social services also have the political support of public employees and their organizations. Furthermore, cutbacks in federal research and information services make it less likely that localities will have access to the accurate information they need to design new building codes or informational programs.

The above discussion suggests several propositions that should be examined to learn from recent experience: 1. The degree of local energy action depends as much on a community's ability to mobilize resources—either locally or through grantsmanship, political connections, and so forth—as on the level of demand for change in local energy conditions. 2. A shortage of outside support will tend to prevent local action on expensive or technically sophisticated energy projects, regardless of their potential payoff. 3. Because outside funds allow a range of energy activities to be treated as distributive, a reduction of outside resources may most seriously affect programs aimed at low-income groups. 4. Technical expertise is more likely to be accepted when it comes from within a community; when it comes from outside, it is more likely to generate conflict or to become an object of blame when problems arise. 5. Technical expertise from outside is better accepted and more fully used when it is used to build local expertise rather than simply to offer expert judgment.

Maintaining Policies and Programs

Like other public actions at the local level, local energy actions are subject to politics. As a result, they can continue to exist in one of two ways. Either they are continually ratified in an open and sometimes exhausting political process, as described in the section above on "Mobilizing Action," or they become institutionalized—that is, they find a stable place in some organizational structure and are treated as ordinary responsibilities of that organization. Institutionalization is evidenced in permanent staff positions, routinized tasks within an organization, and a place in the budgeting and decision-making processes of an organization. Stable, single-purpose physical structures also help institutionalize the activities that take place in them. A resource recovery program is institutionalized both in the facility that processes the wastes and in the position of plant manager, and is usually located in a stable part of the city government, such as the department of sanitation. When a city government adds an energy coordinator, this tends to institutionalize a concern with energy issues. The stability of this concern varies, however, with the permanence of the position, its place in the organizational structure, and its ties to other, long-standing municipal agencies (see Chapter 5).

Institutionalization is important because it maintains the life of a new program with minimal effort. It also facilitates the transmission of skills and knowledge, including political skills and contacts, so that knowledge is not lost when the original staff departs. Institutionalization also implies the development of routines of operation. Such routines make a new unit more efficient in doing its usual tasks, but may also mean the organization is less likely to take risks or promote change. For local energy groups whose goals entail social change, then, institutionalization has drawbacks. There may be a tension between "social goals" and "energy goals."

A plausible proposition to examine is that labor-intensive energy actions may be the most difficult to institutionalize. In the present climate of tight resources, this proposition would mean such actions are especially vulnerable to being discontinued because they tend to draw their labor from volunteers, participants in job training programs, and other unstable sources. When energy action organizations are operated by grassroots community groups, they often start with seed money from an outside source and lack a reliable continuing source of operating funds. And when they are the responsibility of local government, they must be debated in every annual budget. Further, if they fail to find a stable home in a local agency, success in these debates requires continual mobilization of public support.

Once an energy program is operating, it needs different resources, skills, and political support from those that were necessary to get the program started. Budgetary needs tend to stabilize at some level, and needs for technical and managerial skill may replace needs for political organizing ability. Consequently, local community groups may be poorly suited to manage the programs they have helped bring into existence. Although they may be able to mobilize labor and political support, the need to simultaneously develop skills, raise money, and create institutional stability may prove too much of a bootstrap operation for them to manage (Bowden and Kreinberg, 1981). However, such groups may be valuable and flourish as partners with other groups in jointly managed programs.

For local governmental agencies operating energy programs, there are problems of coordination, which emerge most clearly when agencies with different objectives are required to cooperate. For example, when an agency charged with improving the energy efficiency of housing must use the services of trainees in a job training program, the objectives of training and of high-quality work can conflict. This kind of conflict has been a problem for some local energy activities (Stern et al., 1981). In addition, job training programs inherently offer an unstable supply of labor for weatherization efforts (Office of Technology Assessment, 1980). Coordination also tends to be difficult because most bureaucracies offer officials little or no reward for the required behavior (Blumstein et al., 1980). A plausible proposition is that local groups will do best when they have had experience with similar activities in the past. It might be relatively easy for a city agency that had operated heating plants for municipal buildings to manage a new district heating operation. Similarly, a change in the building code may be a fairly routine activity for building inspectors.

It may be especially difficult to implement programs that depend for success on communication and trust. A good example is energy audit and weatherization programs for residences and small commercial buildings (see Chapter 4). For such programs, the local institutions with the resources to initiate programs—chiefly energy utilities and city governments—often find it difficult to involve their intended clientele. Mistrust of these institutions is common and their typical lines of communication in print and mass media tend not to reach poor, old, and non-English-speaking people.[8]

Spreading Energy Ideas

If local energy action is to become a significant national phenomenon, new ideas must spread from where they began to other localities where they might be useful and they must be adapted effectively in those localities. Research should carefully examine the ways local energy ideas have been spread.

Three types of institutions are probably important in spreading ideas for local energy action. First, the mass media transform local events into news stories, thus spreading rudimentary information about new energy ideas and identifying individuals as sources of further information. Media coverage has drawn national attention to such sites of energy innovation as Davis, California; Seattle, Washington; and Fitchburg, Massachusetts.

Second, information can also be spread by specialized institutions, such as state energy offices and the energy programs of the National Governors' Association and the Conference on Alternative State and Local Policies. The same function has been served by the federally funded regional solar energy offices and the National Center for Appropriate Technology, and by such nongovernmental sources as the Center for Renewable Resources, the Institute for Local Self-Reliance, and a number of regional, appropriate technology organizations.

Third, energy information is also spread by various kinds of informal social networks. Extensive research on the diffusion of innovations (e.g., Rogers with Shoemaker, 1971; Leonard-Barton and Rogers, 1981) demonstrates that informal communication with peers is a highly effective medium for spreading ideas—possibly the most effective. This finding seems to hold true with energy ideas (Darley and Beniger, 1981; Leonard-Barton, 1980).

Many social networks have probably been involved in spreading energy ideas, including: national and regional associations of city officials, such as mayors, planners, and engineers; professional associations in architecture, engineering, and the building and transportation industries; and other groups. National environmental organizations, poor-people's organizations, charitable organizations, and the like also spread those energy ideas that may further their organizations' central goals. Personal contacts in such organizations, as well as informal family, friendship, or occupational networks, are also likely to spread energy ideas. New ideas may also spread when individuals who have worked on energy issues in national organizations carry ideas to the areas where they live.

Since spreading energy ideas may have little benefit for the communities that generate the ideas, federal involvement is probably crucial for this function. Past federal actions have probably been significant by establishing and maintaining national and regional energy organizations and by supporting the informal networks of personal interaction that are so important to local innovators. Cutbacks in federal funds since 1980 have meant curtailment of programs that kept local groups in communication with each other and that provided them, at little cost, the benefit of each other's experience. The impact of federal support and its withdrawal on the exchange of energy ideas among communities is worthy of careful study, because local sources are most unlikely to replace a national commitment and because small changes in federal involvement are likely to make a great difference.

IMPLICATIONS AND RECOMMENDATIONS

Knowledge about the phenomenon of local energy activity is severely limited. Because of this lack, we have not attempted to draw any conclusions, but rather to state some plausible propositions about local energy activities that are worthy of careful study. We cannot overstress, however, that these are propositions for testing and not established generalizations. They are based on analysis, judgment, and broad extrapolation from very limited data. But because local energy action can have major positive effects if it can be achieved on a wide scale, we believe that careful research on the identified issues could provide useful insights for national and local officials and for the public.

A simple call for more research is inappropriate, however, because of the nature of the phenomena. We have already noted the great diversity of local energy programs and policies and of the local conditions in which they occur. Because many variables simultaneously influence the way local energy actions emerge and develop, no ordinary program of evaluation research has much hope of isolating and controlling enough variables to reach any definite conclusions, even about a single program. Furthermore, because of the diversity of local conditions, we question how useful any generalizations from research might be to those responsible for making a particular local program work. For the practical purposes of learning from the experiences of other local energy actions, something more than a collection of evaluation studies seems appropriate.

A research effort might begin with careful case studies of existing local energy activities—both successes and failures. Such case studies would have to do much more than simply assess the effects of a program on local energy use. Knowing only that a program succeeded or failed offers little guidance for a new situation. To learn from the experience, it is necessary to clarify the social processes that facilitated, impeded, or transformed a proposed local energy action. From such research, social scientists and practitioners would be able to develop a qualitative knowledge of the processes involved in local energy action—something like the propositions we suggest, but with more detail and a stronger basis in fact.

Such an effort would be a beginning, but it would not be sufficient because local energy activities develop in a context that is constantly changing. Even if a set of generalizations or principles seemed to apply to a local situation at the outset, unforeseen changes in the situation are almost guaranteed. To succeed, a program must be capable of learning from its own experience and the experiences of others. Research results inevitably lag behind such a continuing process. Learning from experience requires more than knowledge about the characteristics of successful and unsuccessful programs. If the goal is to develop adaptive forms of organization, it would be helpful to conduct research on the process by which local energy activities adapt to changing circumstances. In fact, basic research on the general problem of adaptation and learning in organizations might be enlightening. But the development and application of basic research takes time, and local energy activities have immediate needs for knowledge.

One way local groups can learn from each other's experience is by communicating through formal or informal channels. The same communications networks that spread ideas for starting programs may prove useful for spreading ideas about how to solve problems, particularly immediate problems. If national policy is to promote local solutions to energy problems, it would be very helpful to develop ways for localities to learn from each other. This might be done, for example, by creating institutions to share information among communities, by holding conferences, and by making research expertise available to localities so they can carry on continuing assessments of their programs for learning purposes. Though we believe that such techniques would help social programs function better, we suggest them only as experiments: like the local activities themselves, they should be studied as a new social institution so that the society can learn from the experience.

We conclude with a reference to the high hopes many observers have for local action as a way to solve national energy problems. In fact, these hopes concern more than energy: because of the difficulties of past federal programs, many observers want to find ways that the society's response to national needs may be organized collectively at levels below the federal government. Local energy actions provide numerous illustrative examples. It is particularly important at this time to try to learn from the experience of the thousands of recent local energy efforts. Local energy actions provide knowledge not only about the prospects for local solutions to energy problems, but also about what happens when communities try to act on the many different philosophical premises that underlie their particular local energy activities: preserving natural resources; increasing local control over essential resources; redistributing political power and social services; developing the community economic base; or promoting technologies based on appropriateness to the size of the community. If viewed this way, the experiences of local energy actions may provide a way to assess the more general potential for solving national problems at the local level.

The federal government, private foundations, or other interested parties should sponsor research to evaluate the success or failure of existing local energy activities.   Such research should define success and failure broadly because of the close ties these activities have to other local issues. Because the impetus for many local energy actions has come from the redistributional possibilities believed to exist in such actions, the research should explicitly examine the distribution of effort and of benefit from energy programs among the major groups in the communities involved.

The federal government, private foundations, or other interested parties should sponsor research aimed at understanding the processes by which local energy activities begin, succeed, fail, or become transformed.   Such a research program should look broadly at the antecedents and effects of local energy activities. The results of such research would be useful to local groups and for informing national policy.

The federal government, private foundations, or other interested parties should develop and sponsor mechanisms by which localities can learn from each other's experiences with local energy action.   These mechanisms might include holding conferences, making research expertise available to localities to assess their own activities, and supporting travel and communication among local energy officials.

Notes

1. This information comes from D. Streb, United Neighbors of Cleghorn, Fitchburg, Mass., 1982.
2. Local energy actions have been advocated for many reasons, reflecting different views of energy and energy policy. For example, local programs for residential energy conservation have often been advocated as means to meet social needs—to create needed jobs and to provide better housing—as well as for energy services.

Energy actions are sometimes advocated as means to keep money and jobs in the local community, essentially a strategic view at the local level. They are promoted as supportive of environmental values. And in addition, they are defended as good in themselves because they are believed to promote democratic control. For the diverse rationales of a sampling of local conservation programs, see Stern, Black, and Elworth (1981).

3. It has been argued that it is easier to form local coalitions on energy issues than on other social issues. This argument seems plausible because energy does not involve the same sorts of basic social division that marked the civil rights movement and the women's movement in past decades; people who are disadvantaged by energy events do not have daily contact with some clearly definable social group that is not disadvantaged and that might be seen as responsible for their difficulties. Thus, dialogue and coalitions may be more possible across social groups. But the same factors that may make it easier to form coalitions detract from the motivation that may be necessary to mobilize for local action.

4. Information from V. P. Ludlam, Director, Westside Department Corporation, 1981.

5. Information from E. Hagens, Governors State University, Park Forest South, Ill., 1982.

6. Information from J. Pitts, Richmond, Indiana, City Energy Agency, 1982.

7. See note 1 above.

8. For a detailed discussion of these issues, see Stern et al. (1981: Chapter 2).

8

Conclusions and Recommendations

THE ENERGY SYSTEM AND ENERGY POLICY

This book emphasizes certain characteristics of the U.S. energy system and the major actors within it that, we believe, have been given insufficient consideration in energy policy analyses. We believe that fuller consideration of those characteristics will lead to broader and more enlightened policy debates and more effective energy policy. This chapter highlights those characteristics, identifies their general implications for energy policy, and presents some recommendations for action.

Three characteristics of the national energy system have repeatedly impressed us as central: *diversity; uncertainty and mistrust*; and the issue of *control*. Evidence of *diversity* in the U.S. energy system is readily apparent; it is evident even in the ways people and experts think about energy (as we discussed in Chapter 2). Tremendous variation also exists in the needs and practices of energy users, so that analyses based on an average situation are likely to be wrong in many or most particular cases. The effects of diversity may be magnified in an energy emergency, because a varied pattern of shortages may be superimposed on a variety of needs. In addition, perceptions of an emergency vary greatly, even among similarly situated observers. Because of the diversity in the system, decentralized solutions to energy problems must be seriously considered. Our interest in local energy activities (Chapter 7) came partly from the awareness that these activities have been, and can be, field tests of the practicality of using decentralized approaches to a range of energy problems.

Uncertainty affects both the public and policy makers. In an uncertain environment, conflicting interests and changing political positions generate a welter of confusing and contradictory authoritative statements about energy. Consequently, it is no surprise that *mistrust* is endemic. Because energy users are justifiably skeptical of information offered them, the credibility of an information source may make more of a difference than the accuracy of the information offered. Experience leads us to believe that mistrust would be even more serious in a major energy emergency. In short, credibility may be the central problem for energy conservation programs, for emergency preparedness and management policies, and for any other energy programs and policies that rely for their effectiveness on public response.

Energy debates are, in an important sense, debates about *control*. Energy is often symbolic of control or loss of freedom. A major motive behind some local energy activities has been to gain for communities a greater ability to control or manage their destinies, as they see themselves buffeted by national policies and world and national economic forces. The perception and experience of control are also powerful determinants of individual response, and they increase behavioral commitment to future action. Thus, energy policies and programs are likely to be better accepted and more effective if they increase individual and local control, rather than impose decisions from outside. We believe control is currently a particularly important issue in energy policy because of the relatively low level of public trust in energy institutions.

In addition to these characteristics of the energy system, several facts about the actors in the system have also repeatedly impressed us. Although it is often useful to think of these actors as economically rational decision makers, several other processes govern their behavior as well.

One of these processes is *behavioral momentum*. Individuals and organizations are, in part, creatures of habit. They establish routines and stick to them, they work to reduce uncertainty and change in their environments, and they avoid or ignore problems. People, organizations, and local governments often persist in outmoded energy-using practices despite information that they would benefit from change. Individuals, furthermore, usually justify their past behavior to themselves, strengthening their tendency to keep doing what they have done in the past. Organizations create subunits and standard operating procedures that have a similar effect. Behavioral momentum can be reversed by getting the relevant people actively involved in a change process: any new activity tends to commit them to a new course of action. The principle of involvement can be used manipulatively, but it can also be an extension of a democratic decision process: a decision made after political debate constitutes a behavioral commitment for a community.

Another important process is *imitation*. The behavior of individuals and organizations is influenced by example, especially the example of their peers and the opinion leaders of their peer groups. Individuals often follow the examples set by neighbors, relatives, and people they know from work or religious groups; firms follow their competitors or industry leaders; and local governments follow neighboring governments. New energy-saving practices spread through social networks along lines of personal communication, and with knowledge of this process, policy makers can help spread such practices. In energy emergencies, adaptive responses can spread rapidly among communities if there are open communication lines. And the significance of local energy action in the national picture depends critically on how quickly and effectively ideas from one community can be imitated and adapted by others.

A third critical social process is *communication*. Rather than being active seekers of useful knowledge, people and organizations are selective in attending to and assimilating information. Considerable research exists defining the ways this selection occurs. For example, people are more likely to remember and use information when it is presented attractively; when vivid, personalized examples are used; when it is specific to the user's needs; when it comes from a person who is similar to the person receiving it; and when it is presented in familiar and understandable terms. Programs offering information about ways to save energy have not generally made use of such knowledge about communication, and using it could improve their effectiveness. Information designed for emergency use could also become more effective by incorporating what is known about successful communication principles, and especially by making two-way communication possible.

Finally, beyond the characteristics of the system and the processes that affect the actors in it, it must be remembered that individuals, organizations, and political groups have energy-related *values* that affect their actions. We found the expression of values most explicit in the context of local energy action. Values can also be important for conservation policies and programs, because such initiatives are likely to be more effective when they are presented in terms congruent with the values of the intended clientele. Hence, it is necessary to understand the clientele of energy programs if the programs are to be effective.

These characteristics of the energy system and of the key actors within it are often overlooked in policy analysis. Recognition of them leads to some new ways of thinking about energy policy: that policy problems and solutions derive not from single causes but from the complex environments or systems in which action occurs; that policy might aim at making the energy system more adaptable rather than developing specific plans; and that for energy policies to be effective, more careful attention must be paid to the process of developing programs and policies.

Attention to Social Systems

A view of energy policy problems in terms of social systems is implicit throughout this report. This view is best exemplified in three areas.

Energy Information. People get information about energy from many sources: government reports and pronouncements, news stories, energy suppliers, salespeople, friends and acquaintances, personal observation, and elsewhere. A pattern of contradictory information is often likely to come from these varied sources (see Chapter 3), because the value of information often depends on unforeseeable future events and because the information sources often have sizable resources for promoting their conflicting interests and viewpoints. Even if a government effort produced the most accurate available information, it would not necessarily be believed.

A central objective of any government effort regarding energy information must be to design policy and programs to allow skeptics to resolve their doubts. It is therefore important to make information for energy consumers available from personal observation, informal social networks, and other highly trusted sources. In discussing energy emergencies, much of the information normally available will prove irrelevant, and there may be no good way to know in advance what information will be needed or how best to get it. One way to deal with this situation is to keep communication channels open in an emergency, so that those who need information will be able to find it once they define what they need to know. Another approach is to build in ways that official information can be independently verified, because levels of trust in an energy emergency may be even lower than they are under normal conditions.

In all these cases, the public interest is to create a situation that allows good information to be generated and located. The primary objective is for people to be well-informed, rather than for government to produce good information. Sometimes, government may be in the best position to generate the information because it has both the resources and a concern for the public interest, but, even then, government may not be best for distributing the information. In other instances, as with energy-use feedback systems, the objective is for people to generate their own information because it is more meaningful and credible. Here, the government role might be limited to research on feedback devices or methods of making the information clear.

Emergency Response. Contingency plans for energy emergencies are likely to remain unused for many reasons: the emergency that occurs may not be one that was planned for; the responsible officials may think that the situation is different enough so that the plans should be disregarded; or the officials in charge when the emergency occurs may not have been involved in planning and may not learn that a plan exists or find the time to study it (see Chapter 6). Rather than developing detailed plans, therefore, it often makes sense to design institutional systems that can respond effectively to a variety of conceivable emergency situations. One suggestion for system design is to make emergency preparedness a continuing process, so that many people are involved in planning, get practice communicating with those they would need to interact with in an emergency, and gain a sense of competence in dealing with emergencies. All these outcomes should be useful for a wide range of emergency situations.

Local Energy Action.   Local energy action is both a social and political process in individual communities and a national phenomenon. Local action does not arise only from within communities; it is also influenced by ideas that come from outside by a process of social diffusion. In that sense, local actions are influenced by a national communication system consisting of individuals, organizations, publications, and so forth. If it is national policy to encourage local initiatives in solving energy problems, there are many ways to do it. In addition to the more traditional strategies of offering funds and mandating actions, the federal government could support the communication system by holding conferences or assisting particular groups that transmit energy ideas among communities. Such communication may have been a major effect of past federal activities in conservation and solar energy, but it has never been a central concern of policy.

While we offer several system-oriented policy suggestions, we are not confident that each will be the best approach possible. What is important at present is that some new and promising policy ideas can come from thinking about energy problems as the problems of social systems. In a world in which the limits of control by the central government are becoming increasingly apparent, the suggestions that arise from such a perspective may be especially attractive.

Adaptability as an Alternative to Planning

At several points in the discussion of energy emergencies (Chapter 6), we argue that it is important to make the energy system more adaptable so that sudden changes in energy supplies will be less likely to create crises. The goal of adaptability has implications not only for emergency preparedness, but also for energy policy under ordinary conditions.

Adaptability as Part of Emergency Preparedness.   A high priority in policy for energy emergencies is maintaining lines of communication so that decision makers have quick access to whatever information they might need and so that coordinated responses can be developed even for unanticipated problems. We propose that government may need to assure the public that essential needs would be met before certain policy options can be credible. An example of this would be letting price increases allocate fuel in an acute shortage only after ensuring that the basic needs of individuals and communities would be met. We also discuss prevention strategies that would transform parts of the energy system so that a sudden shortage of any single energy source would be less of a problem. These strategies might include broadening the range of fuels used for essential purposes, increasing capabilities for fuel-switching, stockpiling mass transit capability, and encouraging structural change in patterns of building development to decrease the need for fuel for essential travel, such as between home and work. All these approaches emphasize making the energy system adaptable, so that it can find effective responses when needed, rather than devising plans in advance for situations that cannot be precisely foreseen.

Increasing adaptability is a useful strategy for several reasons. First, plans for energy emergencies are often unused when an emergency arises. Second, the existence of several decision options makes it more likely that an appropriate one will be available. Third, because the effects of an emergency are more evident to decentralized actors than to central authorities, it is usually adaptive to decentralize many decisions in a crisis. It must be emphasized, of course, that local authorities need sufficient resources if they are to exercise responsibilities effectively.

Adaptability as a General Policy Strategy. The need for adaptability in energy emergencies has implications for energy policy during nonemergency times. For example, the ability of an energy system to adapt to acute shortage depends in part on its ability to curtail demand quickly, which, in turn, depends on the preexisting pattern of energy use. Different ways of cutting demand in response to nonemergency price increases have different implications for adaptability. In residences, saving energy by curtailment of heating and cooling takes away much of the "slack" that residents have for quick response in an emergency. Saving the same amount of energy by upgrading a furnace or insulating an attic leaves the resident with more capability for quick response. Similarly, gasoline saved by decreasing travel or by ride sharing takes up more of the slack than the same savings achieved by driving one's usual mileage in a more fuel-efficient automobile. Policy regarding energy use and conservation affects adaptability in emergencies; conservation policies that promote energy efficiency may increase adaptability, while those that result in curtailment of amenities can decrease it. Thus far most of the energy savings achieved in the residential sector in response to price increases have come from temperature setbacks and other curtailments, rather than from increased energy efficiency in building shells, furnaces, appliances, and so forth (see Chapter 4). Hence, a more vigorous effort for residential energy efficiency, rather than an emphasis on "conservation" defined as unspecified energy savings, might be a positive contribution to emergency preparedness.

Adaptability also implies that diversifying the national energy production and distribution system would help prevent emergencies. To the extent that communities, regions, and the nation depend on a broader range of fuels, they are less vulnerable to a disruption in the supply of any one. To the extent that distribution systems for oil, electricity, natural gas, and renewable energy supplies are decentralized, single accidents or hostile acts will have less effect on the entire system. Thus, certain political and economic decisions about the structure of energy production will have significant impact on the society's adaptability in the face of problems in the energy supply system.

For a national policy that emphasizes adaptability, certain kinds of local energy management are relevant. For example, development of locally available energy sources tends to broaden the national mix of energy supplies, therefore increasing adaptability. In the long run, local zoning policy can increase adaptability by making urban life less vehicle-dependent. And the knowledge and experience that come from local energy activity may leave local decision makers better skilled and better informed when the time comes for them to make quick responses under stress. If the national strategy for energy emergencies stresses adaptability, the federal government may have an interest in encouraging some types of local energy activities.

Broader Implications of Adaptability.   Increased adaptability implies increased control by local rather than central decision makers, and it implies a focus on energy efficiency rather than unspecified energy conservation. In these respects, the idea of adaptability relates to several of the themes that have recurred in our study, and to a central policy debate of the 1980s.

The argument for adaptability in emergencies essentially derives from a recognition of diversity in the energy system and a realization that in times of rapid change, and especially in an acute supply shortage, central authorities are in a poor position to assess diverse local conditions and meet local needs. In short, the argument for adaptability draws on the themes of diversity and uncertainty. It is also related to the themes of trust and control. Plans for emergencies that allow local decision makers to check on the information they get and to make their own decisions would certainly increase local control, and might also increase people's trust in the information and advice they are offered.

Adaptability, with its connotations of local self-determination, seems an attractive idea for the 1980s, given the recent history of energy policy.

In the mid–1970s, changing world events brought about general agreement in the U.S. that an expanded federal role in energy decision making was needed. Though consensus was lacking on the goals of national policy, it was agreed more incentives were needed for energy conservation and production. The federal government moved into energy policy, as it had expanded into other policy arenas in the previous decades. But as with other areas of policy, tight federal control of local activities sometimes led to widespread disenchantment with federal policy. Probably the worst instance was the 1979 gasoline shortage, when federal allocation plans were accused of making matters worse. There was the sense that private, state, or local decisions might have been more effective and more appropriate than federal control.

In this context, the connection of local control to the adaptability of the national energy system feeds into recent calls from across the political spectrum for more local control and for less outside interference. While local control seems desirable as an ideal, we do not wish to make a general endorsement of an unspecified principle of local control: the ability of local groups to manage energy remains largely unknown (see Chapter 7). Also, there may be fairly serious limits to the range of energy options local groups can pursue, especially in the absence of sources of outside expertise, money, and ideas. More experience making local energy management work is needed before major responsibility for energy adaptability can be delegated to local institutions.

To increase the adaptability of the national energy system by emphasizing local control requires the development of new relationships between the federal and local levels. It may be useful for national policy makers to think of themselves as facilitating the operation of somewhat self-regulating social systems, rather than as planning or controlling energy activities throughout the nation. The notion of federal support for systems of communication among localities is one of several examples of such thinking. It may also be useful to think of policies embodying new federal/local relationships as experimental trials in the sense that while they may not succeed at first, they may evolve through a learning process toward a relationship in which both local and national levels contribute, with net benefits both for local control and national flexibility. If this happens, effective partnerships will evolve through a political process in which both local and national interests are represented.

Experimentation: Finding Out What Works

Many of the conclusions reached in this report are rather general, at least compared with what is needed to draft a detailed piece of legislation or a government regulation. This may be frustrating to a reader wishing to develop or modify specific programs based on some of the ideas in this book. We have refrained from being more specific more often because while we have confidence in the principles outlined here, we do not yet know what will work in particular situations. We believe it is usually not too difficult to find out what works, but that it is prudent not to overstep our knowledge. This section describes our perspective on the process of developing effective energy policies and programs.

All new government policies and programs are, in effect, experiments. To treat new policies as anything more than experimental is to set the public up for disappointment, for few policies turn out exactly as expected. Well-meaning efforts by government and industry to make things better often produce unexpected effects—and sometimes these effects are so undesirable that many people feel they would have been better off if no effort had been made. We do not believe, however, that the solution is to do nothing, because it is impossible to do nothing. If government were to discontinue its efforts to improve housing or education, for example, housing and education would continue, but differently. Government retrenchments are experiments, too. And because of the complexity of society, it is difficult to foresee the results.

In some areas of energy policy, the experimental nature of innovations is recognized. For example, decision makers would not dream of introducing a new energy technology without careful laboratory and field testing. There are fairly well-established procedures for this process of research, development, and demonstration. Basic research is done on the physical, chemical, biological, or other processes involved, and a prototype is built, tested, and carefully observed. If the technology still appears promising, large or mass-producible prototypes are constructed, and then a production process of commercial size is tested. Eventually, after this exhaustive testing, the technology is observed under naturalistic conditions, and it goes into commercial operation. During all stages, careful measurements of energy transfers, inputs, effluents, and costs are made, so that decisions about technical and economic feasibility can be based on accurate knowledge about the performance of the technology under increasingly realistic conditions.

This careful procedure contrasts sharply with the usual method of introducing conservation and information programs, tax credits, price decontrol, and other energy policies. Such social and economic policies are usually established by legislation or administrative decision and most often are implemented in toto and at once, with minimal or no observational knowledge of how the policies may actually work. On those few occasions when field testing is done, the quality of evaluation is poor, leaving open as many questions as it answers about how or whether the policy will be effective (e.g., General Accounting Office, 1981). From our perspective, this practice amounts to a policy of experimentation on human populations with insufficiently tested techniques. Worse, the practice occurs in spite of the existence of a variety of methods for finding out what works that are both effective and respectful of the interests of participants.

In medical policy, for example, there are procedures for introducing new drugs that specify laboratory trials *in vitro* and in animals before human trials begin. Only if a drug is promising is it tried on small groups of carefully selected human subjects. Society has created procedures of peer review and informed consent to protect the interests of human subjects, and it has devised regulatory procedures to block ineffective products from reaching the market.

Effective methods for conducting field trials also have been developed for programs or policies for which animal research is impossible. There is a vast literature on methods of evaluating the effects of programs in the field (e.g., Campbell and Stanley, 1971), which we will not attempt to summarize here. Often, these methods involve statistical matching of population or individual characteristics. More complete scientific control can be achieved by randomly assigning participants either to the program being tested or to a group that will be exposed to the program at a later date, a "waiting-list control." This is done so that the effects of the program can be disentangled from those of unrelated events that may occur during the research.

The best scientific procedure, of course, is controlled experimentation, in which participants are randomly assigned to a range of program alternatives simultaneously and in which they may be monitored before, during, and after exposure to the program. While it sometimes seems impractical or unethical to expose human populations to such a procedure, it has occasionally been done with success and without raising any serious objections. Such a procedure requires that both scientific and ethical issues be approached carefully. An example is the Wisconsin time-of-use electricity pricing experiment (Black, 1979). The researchers were reluctant to use only volunteer participants because the study might attract only those households that already used most of their electricity in off-peak periods and that would automatically benefit from the experimental rates. The behavior of such experimental subjects would have led to incorrect interpretations about how people, in general, would respond. But there was an ethical problem with simply assigning customers to the new rates: the experiment would increase electric bills for some of them. A "hold-harmless" procedure, in which the rate structure would be set low enough to prevent that outcome, was considered an unrealistic test and was rejected.

The researchers' solution was to convene random samples of people to act as juries on the question of what rates would be fair to the participants in the field trial. The juries agreed that it would be fair to set rates so that the average consumer would experience no change in bills if peak electricity use remained unchanged. While this meant that participants who normally used a higher-than-average proportion of their electricity in peak periods would have to pay higher bills unless they shifted electricity use to off-peak hours, the juries considered this fair. When the rates were explained to the participants, they also considered them fair.

As this example shows, with ingenuity, it is often possible to use field trials to gather strong evidence about the effectiveness of programs, while avoiding serious ethical problems. It is possible to devise procedures analogous to those used in medicine to test new program ideas first in tightly controlled small trials involving only scores or hundreds of participants. Statistical techniques can give a rough determination of a program's effectiveness on the population from which those participants come. If the program looks promising, it can then—or simultaneously—be tested in other small field trials on different populations, to see if the program works equally well under diverse conditions. With the knowledge gained from these trials, decisions can be made about how or whether to proceed to more extensive trials or to actual policy implementation.

This sort of procedure may seem expensive and time consuming. But while field trials do take time, they are often less expensive than survey research projects, which are limited because they can either only gather people's reactions to hypothetical policies or be used after a policy has been widely implemented. And the costs of not conducting field trials can be enormous. Probably the clearest recent example, although not in the area of energy, is the introduction of the Susan B. Anthony dollar. Even very limited testing of prototypes in field settings—with cashiers, in vending machines, at toll booths, and so on—would quickly have established the need to redesign the coin. Even simulation experiments in settings created to imitate real situations might have been sufficient. Instead, the unacceptable coin was introduced nationwide at great expense, and it will now probably be many years before a new dollar coin is introduced, even though inflation and the short stressful life of paper currency may make it more and more imperative that dollar coins enter circulation. A similar failure of government to pretest new policies and programs characterizes the energy field. For example, federal and state conservation and solar tax credits were introduced—without field testing—to encourage investments that would otherwise not have been made. However, it was also possible, as some analysts argued, that the credits would merely provide a windfall for people who would have invested anyway (e.g., Rodberg and Schachter, 1980). If the tax credits mainly produce a windfall, the "experiment" will have failed and will have been very costly for taxpayers.

For all the above reasons, new policies should be recognized as experimental, and the best appropriate research methods should be used in field trials of new policies and programs before they are generally introduced. Given the diversity in the energy system, it will usually be important for several field trials to be conducted, to learn what happens under a variety of conditions. Statistical techniques often allow more to be learned from several small trials than from a single, larger trial.

Rigorous evaluation methods should be part of the policy development process. That is, research done in an early stage of policy development can be used to modify a prospective policy to make it more effective or to make it more acceptable to the people it is meant to serve. In the jargon of evaluation researchers, the process is called "formative evaluation": evaluation of prototypes conducted for the purpose of formulating the policy or program. The alternative, "summative evaluation," is used to judge the success of a program after the fact and is not intended as a tool for redesign.

In our judgment, it is a waste of resources to conduct field tests and evaluations of energy policies and programs merely to judge the projects against some criterion of efficacy, cost-benefit, or the like. While overall outcome measures sometimes have a useful purpose, it must be remembered that the purpose of a field trial is to improve future policies and programs, not just to judge the past. Observation should be aimed at aiding the learning process within a project and at gaining understanding of how the project can be adapted to other conditions. It is important to recognize that programs and policies in the United States, with its complex political system, are never rigid, nor should they be. Ideally, people and institutions learn, and programs improve. So while tight experimental controls are appropriate and necessary to prove the concepts involved in energy policies, an operating program should not be held to a set of rigid procedures, which would transform the program from a vital, changing social activity into an unrealistic laboratory-style experiment. The overall goal should be to develop programs that are relatively self-regulating, rather than to find an ideal set of procedures. It is therefore important to monitor the processes of change in an operating program to see if they are effective and responsive, if the program tends to become isolated from the people it was intended to affect, or if some other sort of transformation occurs.

It is also important to monitor the effects of energy programs and policies on other parts of the social system. Given society's experience with undesirable secondary effects of some past programs and policies, people are understandably wary about new public activities. Evaluations of new programs and policies should therefore avoid being too narrowly focused on single objectives. However, a broad evaluation can be accomplished only imperfectly by social scientists because not enough is understood about social systems to be able to identify with confidence all the secondary effects that may result from a given activity. It is advisable to involve representatives of groups likely to be affected by a policy or program in the evaluation process from the start—to ensure that the concerns of the groups are addressed in the evaluation and to make the evaluation more credible to the people it is intended to benefit.

New Roles for Major Social Institutions

It is important to realize that an emphasis on adaptable social systems suggests different roles for the federal government than those it has normally played. It also suggests different and often new roles for local institutions and for the scientific community. The federal government has had several roles in energy policy: it has been a source of money, it has conducted technological research, it has attempted direct control of activities through regulation, it has collected and distributed information, and, more recently, it has backed away from many of these activities. In all these roles, the federal government either asserts control—through the use of regulations or categorical grant procedures, for example—or relinquishes control, sometimes providing resources to other institutions, sometimes not.

But other roles are possible, such as those that facilitate action by other institutions: helping localities learn from each other, maintaining communications between local officials so they will be able to interact effectively in an emergency, and so forth. The federal government can serve a linking function in an energy system characterized by decentralized control. This approach offers an alternative to the stereotyped debate over whether it is better to have more or less federal involvement. Of course, this concept of the federal role needs further debate and analysis. While we suggest that a linking role is sometimes appropriate, we have not addressed the question of when this role is more or less appropriate than the traditional federal roles. Neither have we considered in detail what, specifically, federal agencies would do in a linking role.

Increased decision-making power for local groups implies an increased local need for resources and expertise. But what resources and expertise? And where would they come from? We have not explored these questions. Nevertheless, because of the obvious need in the society to consider new definitions of the federal role, we offer for debate and investigation the idea that the federal government can sometimes facilitate local control, rather than either taking control itself or leaving localities essentially on their own to solve their problems.

Our analysis of the energy system and energy policies also implies new roles for the scientific community. Under the approach we have suggested, scientific methods are used to help those involved in a new policy or program learn from their experiences. For scientists, this approach implies closer attention to social processes, rather than evaluation of outcomes. It also implies a willingness of field researchers to trade some degree of experimental control for an understanding more relevant to constantly changing social institutions and to trade some control over defining the question to be studied for an understanding more useful and acceptable to the relevant publics.

These alternative roles for the federal government, local institutions, and scientists will only come into being if the political process determines that an energy system relying more on adaptability is desirable. We realize that there are values and interests on both sides of that issue. For example, depending on the situation, there are various interests served by central planning. Also, interests that benefit from a narrow commodity definition of energy may see the market as an adequate institution for promoting adaptation—which it is, within limits. But as we noted in Chapter 2, the market allows adaptation as a function of an actor's market power. For financially strapped municipalities, for future generations, for the poor, and for preventing major energy emergencies, considerations of market efficiency alone do not necessarily produce the best results for the entire economy and society. Some notions of adaptability require public action in addition to the workings of the market. In short, decisions about the roles of the federal government, the market, and other institutions in energy policy are ultimately political decisions. Our discussion of a possible federal role in facilitating adaptation adds an alternative to the list for public consideration.

This approach may also be useful in other areas in which the government, as in energy policy, operates with serious limitations on its ability to control. Increasing control, designing and modifying social systems, and using scientific methods as an aid to social learning may be relevant to other policy problems that, like energy, are characterized by diversity, uncertainty, and mistrust. Such policy issues seem almost ubiquitous in the 1980s, but the relation of the general concepts discussed here to nonenergy issues is the subject for other investigations.

RECOMMENDATIONS

Our analysis of some major issues in energy policy has produced some new ways of thinking about the problems that have broad policy implications. It is not always obvious, however, how to translate principles into specific policies or programs. It is also not always clear whether the needed action should come from public or private sectors. Here, we offer specific recommendations that we can support with confidence. More detail and research support for these recommendations can be found in previous chapters.

Conservation Programs

1. Government, utilities, and other operators of conservation programs should apply the best existing knowledge of communication processes in presenting information to energy users.   Currently, in the design of utility bills, fuel economy mileage guides, energy efficiency labels, and other forms of conservation information, government and utilities implicitly and incorrectly assume that energy users will automatically notice and understand any information that is available. Instead, the presentation of all conservation-oriented communications must pay careful attention to format, timing, clarity, vividness, framing of information, characteristics of the information source, and other important factors (see Chapter 4 and, with regard to organizations, see Chapter 5).

2. Programs that use energy auditors, extension agents, "house doctors," or other outreach workers should train these personnel in communication skills.   General communication skills—involving paying attention to one's audience, answering questions, and so forth—are easily overlooked in programs that emphasize energy savings through technical improvements, because it is possible to think that the best thing for the energy user to do is to leave the technology alone. In fact, the communicator's skills may influence the adoption of energy-efficient technology, the energy user's future adoption decisions, and those behaviors that affect how efficiently the technology performs. Outreach workers should also be trained in skills specific to the communications they deliver: vivid methods of presentation; effective ways of framing decisions; understandable units of analysis; and so forth.

3. Programs that emphasize energy information should distribute it through sources that are trusted by the intended audience.   Because of the diversity of energy users, very little information is universally applicable. Furthermore, energy users justifiably mistrust many information sources. It is necessary to match information to the energy users for whom it is relevant and useful and to find credible information sources for different groups of energy users. Information for industrial groups is likely to be more effective if it comes from trade or professional associations or specialized trade journals. Within organizations, information will be more effective if it comes from highly placed sources. Information for local officials may have more impact if it comes through their associations or is delivered at their conventions. Information for households may be more effective if the market is segmented to offer information to different groups through different sources: consumer publications for some; community and church groups for others; and so forth.

4. Energy information and building retrofit programs, such as the Residential Conservation Service, should continue to stress consumer protection features.   Such features include strict standards for materials, independent and regular inspection of services, independent mechanisms for handling conflict, and certification of service providers. The available research indicates that fear of inaccurate information or incompetent installation is a significant barrier to residential investments in energy efficiency and that effective assurances of consumer protection can greatly increase levels of investment.

5. Energy audit information should be delivered in person.   Personal contact increases attention, and more effective communication is possible in person. We believe that the 1982 federal decision to eliminate the requirement of in-person communication from the regulations for the Residential Conservation Service was ill-advised. It seems certain that some programs will no longer present audit information in person and, consequently, that consumers will probably take less action on valuable recommendations from energy audits.

6. On-site energy audits and similar outreach programs should give energy users personal experience in making their energy systems more efficient.   Such experience creates a behavioral commitment that makes further action more likely and increases energy users' knowledge of and control over their energy systems.

7. Experiments should be continued with systems that provide incentives for purveyors of energy information and services to offer safe and effective information and thorough work.   One example is the approach being tried in New Jersey in which an organization that offers energy audits is paid *only* for the energy saved. The general principle of built-in incentives is applicable in many areas. For example, in buildings, inspection requirements create incentives for careful work by insulation contractors; in locally managed energy programs, the activity's visibility to its clients provides a built-in incentive, because community members can withhold funds, votes, or participation.

Conservation Policies

8. Federal or private agencies should develop simple, understandable indices of energy efficiency, comparable to miles-per-gallon, for appliances, furnaces, and building shells.   Because of the history of increasing energy invisibility (see Chapter 3), energy is not easily understandable to most people. Often the most accurate units of measure, such as therms-per-degree-day, have little intuitive meaning. Simple indices can attract energy users' attention and can also be useful guides for action, especially if they can be verified against users' experiences. There have been some federal efforts in this direction, which are commendable. Controlled experimentation with alternative indices is probably the best way to determine their effectiveness and usefulness.

9. Labeling, rating, and certification programs should be supported to ensure that indices of energy efficiency come into common use. More energy efficiency is purchased when simple indices become easily available. There have already been some commendable federal and private initiatives toward labeling appliances and vehicles and rating or certifying buildings for energy efficiency. Some uniform system of measurement and some uniform labeling format should exist for each class of goods being certified or labeled. The question of whether such standards should be developed by the federal government or some other source hinges on credibility (see Chapter 4). The decision on whether labeling or certification of energy efficiency should be voluntary or mandatory depends on the extent to which energy-efficient decisions are considered to be a public interest. Past experience indicates that private interest has only occasionally led to certification programs.

10. Public agencies and private firms should develop better ways to give energy users feedback. Techniques should include monitoring devices for use in buildings and vehicles and more informative utility bills. Feedback is necessary because energy flows are generally invisible to energy users; the effects of efforts to cut energy use are also very difficult to judge (see Chapter 3). Feedback provides information that is specific to an energy user's situation, and it makes that information highly credible. Research should be expanded to identify understandable measurement units for feedback. When the units are for residential energy users, the research should be done by the public sector, because it offers some benefits to the general public and because the cost of identifying appropriate units cannot be recovered by individual manufacturers that might do the research. For defining understandable units of energy use for organizational energy consumers, the research might appropriately be done by large organizations or their trade associations. Research should also proceed on attractive feedback displays; this work seems appropriate for the private sector, as it can be incorporated into equipment designs that can compete with each other. Feedback equipment should be demonstrated and tested in pilot projects, supported by public or private funds. More informative utility billing systems should be field tested and considered by regulatory agencies as possible requirements.

11. Government or business groups should collect and disseminate highly condensed summaries of energy conservation efforts by organizations, including benefit-cost analyses. Because attention is a scarce resource for organizational decision makers, brevity is important. And because decision makers understand and value benefit-cost analysis, it may be a particularly effective form of communication. The summaries should be disseminated by sources that get the audience's attention and are seen as credible, such as trade associations. Success stories from the leading organizations in a field should be especially convincing.

12. Trade and professional associations should establish or expand regular networks within which organizational representatives concerned with energy can interact. Such networks should include people whose professions partly concern energy, such as architects, engineers, building managers, and so forth, and also associations in which only some members may take an interest in energy, such as groups of municipal officials, planners, mortgage lenders, industry representatives, and scientists. These networks might take the form of subgroups within an association, but the specifics are less important than the goal of establishing and maintaining communication networks so that good, new ideas and innovative practices can spread.

13. Government or other actors should provide energy information in disaggregated fashion for diverse users. Energy information is not equally relevant or even accurate for all potential uses and users. For energy-efficient improvements for building shells, for example, only disaggregated data can become credible. Data should be gathered from occupied buildings, rather than relying primarily on engineering models. When energy-using products are highly standardized, as with automobiles and appliances, information-gathering can be centralized.

14. The professionalism and independence of federal statistical collection and reporting systems should be protected. These systems represent valuable sources of credible energy information. Several national statistical services have established their credibility in years of service to experts and the public (see Chapter 3). Furthermore, federal agencies are in the best position to collect information that is comparable across regions. Because of the diverse needs and situations of energy users, both under ordinary conditions and in emergencies, it is important that credible information about energy users continue to be collected; that data be collected at a disaggregate level; and that this process be standardized and overseen by a centralized agent. Policies that undermine the effectiveness or integrity of credible federal statistical agencies make it harder to achieve credibility, which will be especially critical in the event of energy emergency.

15. On a trial basis, government and private foundations should provide financial support for local consumer groups and other organizations that will gather and exchange information on consumers' experiences with local purveyors of energy-saving equipment and services. Such organizations, similar in purpose to Consumers Union, which publishes *Consumer Reports,* could enable people to check the quality of services provided at the local level, where national consumer organizations cannot be very helpful. Such organizations, once they establish their credibility by giving useful energy information, could later become self-supporting by monitoring other local consumer services, such as home remodeling and appliance and automobile repair.

Organizational Factors in Energy Use

16. To improve energy efficiency, private organizations and government agencies should build energy concern into organizational action by employing routines such as separate budgeting of energy costs and life-cycle accounting for energy-related capital investments.

17. To produce long-lasting change in energy-using practices, private organizations and local energy programs should entrust to an organizational subunit that has a permanent place the responsibility for monitoring and maintaining the changes. Preferably, this subunit should be highly placed in the organization. Changed practices are more persistent when they are institutionalized.

18. Experiments should be conducted to negotiate agreements to share the costs and benefits of energy efficiency investments between building owners and occupants, especially in multifamily housing. Because such arrangements may not transfer easily, we recommend that pilot projects involve interested parties outside the particular building, such as consumer groups or business associations. Their involvement may help spread the word of a successful agreement and also commit the interested observers to trying to adapt it to other situations.

Energy Emergencies

19. The federal government should develop an organizational structure for energy emergency preparedness that is broadly based, related to a wide variety of possible contingencies, and well suited to provide continuity for the long term. This structure should be linked with complementary structures at the regional and local levels. Because the possible energy emergencies are varied and their local effects very diverse, it is virtually impossible to develop contingency plans for the specific conditions that will arise. Policy for preparedness should emphasize improving the ability of social systems to respond to diverse and even unanticipated emergency conditions.

20. By means of such mechanisms as exercises and emergency planning efforts, the federal government should use its energy emergency preparedness programs to alert key people to potential problems and to train them in certain aspects of crisis behavior, such as the use of emergency communication systems. People are more likely to act on the basis of personal knowledge than on information contained in documents. Since the nature of an emergency will probably be a surprise, it is more important that key actors be familiar with procedures of general utility than with specific actions useful only in a certain type of crisis.

21. Energy emergency preparedness programs should be concerned with post-emergency recovery as well as with crisis management, because recovering effectively from one emergency may be essential in preparing for another.

22. The federal government should assure that key parties in the United States can communicate with each other quickly in the event of an energy emergency. Communication is the most effective way for coping with the diversity and uncertainty of emergency situations; it is also necessary for checking the accuracy of available information and for spreading knowledge of effective ways for coping with the crisis. The starting point is such basic matters as lists of key personnel, both public and private, and notes on how they can be contacted in a hurry, but it is also necessary to provide alternate means of communication in case the normal system is congested or interrupted.

23. The federal government should carefully consider what kinds of information can be "stockpiled" ahead of time for an emergency. For instance, most kinds of detailed information will become obsolete quickly, but prepackaged news releases and television spots that provide examples of coping strategies (for example, travel options if petroleum products are very scarce) could be very useful.

24. The federal government should investigate how to design a public information program that is credible to a broad cross-section of society. Such a program would need to include ways to respond to public concerns and, in some cases, it might need to provide mechanisms for verifying the information provided.

25. The federal government should identify conflicts that might be caused by state and local actions in an emergency, including possible conflicts between net producer and net consumer states, and seek to resolve important issues before an emergency occurs.

26. As part of the effort to prepare for emergencies, the federal government should support research:

    a. to ascertain the potential roles of social groups and relationships in a decentralized approach to emergency preparedness, together with options to enhance their potential;

    b. to learn more about the importance of such factors as leadership, group norms, and media attention in maintaining public support for emergency preparedness;

    c. to identify the critical needs that would have to be met in an energy emergency and alternatives for meeting them; and

    d. to identify the types of information needed during an energy emergency, the best sources for providing that information under emergency conditions, and the likelihood that those needs will be met without federal government initiatives.

Local Energy Actions

27. The federal government, private foundations, or other interested parties should sponsor research to evaluate the success or failure of existing local energy activities.   Such research is necessary to judge the potential of local energy activities to make the national energy system more flexible and resilient and to address the problems caused for energy policy by the diversity of local conditions, the widespread mistrust of energy information and institutions, and the expressed need of people for increased control over their destinies. Evaluation research should define success and failure broadly because of the close ties energy activities have to other local issues and because the purpose of the research is to aid societal learning rather than simply to pass judgment on a program or policy. Because the impetus for many local energy actions has come from the redistributional possibilities believed to exist in such actions, the research should explicitly examine the distribution of effort and of benefit from energy programs among major groups in the communities involved.

28. The federal government, private foundations, or other interested parties should sponsor research aimed at understanding the processes by which local energy activities begin, succeed, fail, or become transformed.   Such a research program should look broadly at the antecedents and effects of local energy activities. Results would be useful to local groups and for informing national policy.

29. The federal government, private foundations, or other interested parties should develop and sponsor mechanisms by which localities can learn from

each other's experiences with local energy action.   Such mechanisms include holding conferences, making research expertise available to localities to assess their own activities, and supporting travel and communication among local energy officials, among others. These mechanisms are needed to facilitate the spread of effective local innovations to other localities.

References

Abrahamse, A., and Morrison, P. A. (1981) Demographic Influences on Energy Use: An Analysis Based on the Panel Study of Income Dynamics. Unpublished draft paper. Santa Monica, Calif.: Rand Corporation.

Acton, J. P., and Mowill, R. S. (1975) Conserving Electricity by Ordinance: A Statistical Analysis. Unpublished report. Santa Monica, Calif.: Rand Corporation.

Adams, R. N. (1975) *Energy and Structure*. Austin: University of Texas Press.

Adams, W., and Dirlam, J. B. (1966) Big steel, invention and innovation. *Quarterly Journal of Economics* 80:67–189.

Allison, G. T. (1971) *Essence of Decision: Explaining the Cuban Missile Crisis*. Boston: Little, Brown.

Altman, E. I. (1971) *Corporate Bankruptcy in America*. Lexington, Mass.: Lexington Books.

Anderson, W. A. (1969) Disaster warning and communication processes in two communities. *Journal of Communication* 19(2):92–104.

Argote, L. (1981) Input Uncertainty and Organizational Coordination in Hospital Emergency Units. Graduate School of Industrial Administration Working Paper #6–8081. Pittsburgh: Carnegie-Mellon University.

Aronson, E. (1969) The theory of cognitive dissonance: a current perspective. In L. Berkowitz, ed., *Advances in Experimental Social Psychology*. Vol. 4. New York: Academic Press.

Aronson, E. (1980) Persuasion via self-justification: large commitments for small rewards. In L. Festinger, ed., *Retrospectives on Social Psychology*. New York: Oxford University Press.

Aronson, E., and Golden, B. (1962) The effect of relevant and irrelevant aspects of communicator credibility on opinion change. *Journal of Personality* 30:135–146.

Aronson, E., and Mills, J. (1959) The effect of severity of initiation on liking for a group. *Journal of Abnormal and Social Psychology* 59:177–181.

Aronson, E., and O'Leary, M. (1983) The relative effectiveness of models and prompts on energy conservation: a field experiment in a shower room. *Journal of Environmental Systems* 12:219–224.

Aronson, E., Turner, J. A., and Carlsmith, M. (1963) Communicator credibility and communication discrepancy as determinants of opinion change. *Journal of Abnormal and Social Psychology* 67:31–36.

Axelrod, R. (1981) Eighteen Dilemmas of Oil Vulnerability. Institute of Public Policy Studies, University of Michigan, February.

Axsom, D., and Cooper, J. (1980) Reducing weight by reducing dissonance: the role of effort justification in inducing weight loss. In E. Aronson, ed., *Reading About the Social Animal.* 3rd ed. San Francisco: Freeman.

Baran, A., Lakonishok, J., and Ofer, A. R. (1980) The information content of general price level adjusted earnings: some empirical evidence. *The Accounting Review* 55: 22–35.

Barton, A. H. (1969) *Communities in Disaster.* Garden City, N.Y.: Anchor-Doubleday.

Baum, A., and Singer, J. E. (1981) *Advances in Environmental Psychology.* Vol. 3. *Energy: Psychological Perspectives.* Hillsdale, N.J.: Lawrence Erlbaum Associates.

Beck, P., Doctors, S. I., and Hammond, P. Y. (1980) *Individual Energy Conservation Behaviors.* Cambridge, Mass.: Oelgeschlager, Gunn & Hain.

Becker, L. J. (1978) The joint effect of feedback and goal setting on performance: a field study of residential energy conservation. *Journal of Applied Psychology* 63:428–433.

Becker, L. J., Seligman, C., and Darley, J. M. (1979) Psychological Strategies to Reduce Energy Consumption. Project summary report prepared for the U.S. Department of Energy. Princeton, N. J.: Center for Energy and Environmental Studies, Princeton University.

Berger, P. L., and Luckman, T. (1966) *The Social Construction of Reality: A Treatise in the Sociology of Knowledge.* Garden City, N.Y.: Doubleday.

Berry, J. M. (1981) Programs in Energy Targeted. *Washington Post,* March 21, p. A1.

Beyea, J., Dutt, G., and Woteki, T. (1978) Critical significance of attics and basements in the energy balance of Twin Rivers Townhouses. In R. H. Socolow, ed., *Saving Energy in the Home: Princeton's Experiments at Twin Rivers.* Cambridge, Mass.: Ballinger.

Black, J. S. (1978) Attitudinal, Normative, and Economic Factors in Early Response to an Energy-Use Field Experiment. Unpublished doctoral dissertation, Department of Sociology, University of Wisconsin.

Black, J. S. (1979) The Role of Social Scientists in Field Experiments on Energy. Paper presented at the meeting of the American Psychological Association, New York, September.

Blair, J. M. (1976) *The Control of Oil.* New York: Pantheon Books.

Blau, P. M. (1964) *Exchange and Power in Social Life.* New York: Wiley.

Bleviss, D. L. (1980) Unpublished proceedings of the Multifamily and Rental Housing Workshop, a conference organized by the Federation of American Scientists Fund, Washington, D.C., December.

Bloom, P. N., and Novelli, W. D. (1981) Problems and challenges in social marketing. Unpublished manuscript. Cambridge, Mass.: Marketing Science Institute.

Blumstein, C., Kreig, B., Schipper, L., and York, C. (1980) Overcoming social and institutional barriers to energy conservation. *Energy* 5:335–371.

Booz, Allen, and Hamilton. (1980) The low cost/no cost energy conservation program in New England: An Evaluation. Prepared for the U. S. Department of Energy.

Borgida, E., and Nisbett, R. (1977) The differential impact of abstract vs. concrete information on decision. *Journal of Applied Social Psychology* 7:258–271.

Boulding, K. (1979) In extremis. *Technology Review,* August/September:8–9.

Bowden, C., and Kreinberg, L. (1981) *Street Signs Chicago.* Chicago: Chicago Review Press.

Brehm, J. W. (1956) Postdecision changes in the desirability of alternatives. *Journal of Abnormal and Social Psychology* 52:384–389.

Brehm, S., and Brehm, J. W. (1981) *Psychological Reactance: A Theory of Freedom and Control.* 2nd ed. New York: Academic Press.

Brewer, G. D. (1973) *Politicians, Bureaucrats, and the Consultant: A Critique of Urban Problem Solving.* New York: Basic Books.

Breznitz, S. (1976) False alarms: their effects on fear and adjustment. In I. Sarason and C. Spielberger, eds., *Stress and Anxiety*. Vol. 3. New York: Wiley.

Britan, G. M. (1979) Evaluating a federal experiment in bureaucratic reform. *Human Organizations* 38:319–324.

Brunner, R. D. (1980) Decentralized energy policies. *Public Policy* 28:71–91.

Brunner, R. D., and W. E. Vivian (1979) Citizen viewpoints on energy policy. Institute of Public Policy Studies, University of Michigan, July.

Bupp, I. C. (1979) The nuclear stalemate. In R. Stobaugh and D. Yergin, eds., *Energy Future*. New York: Random House.

Burton, D. (1980) *The Governance of Energy*. New York: Praeger.

Burton, I., Kates, R., and White, G. (1978) *The Environment as Hazard*. New York: Oxford University Press.

Buss, A. H. (1980) *Self-consciousness and Social Anxiety*. San Francisco: Freeman.

Campbell, D. T., and Stanley, J. C. (1971) *Experimental and Quasi-experimental Designs for Research*. Chicago: Rand McNally.

Carter, C. F., and Williams, B. R. (1959) The characteristics of technically progressive firms. *Journal of Industrial Economics* 7(March):87–104.

Center for Renewable Resources. (1980) *Renewable Resources: A National Catalog of Model Projects*. 4 vols. Washington, D.C.: U. S. Department of Energy.

Chakrabarti, A. K., and Rubenstein, A. H. (1976) Interorganizational transfer of technology: a study of adoption of NASA innovations. IEEE Transaction on Engineering Management EM–23, February:20–34.

Chandler, A., Jr. (1962) *Strategy and Structure: Chapters in the History of the American Industrial Enterprise*. Cambridge, Mass.: MIT Press.

City Currents. (1982) Geothermal growth in Susanville. January.

Cohen, M. D., and March, J. G. (1974) *Leadership and Ambiguity: The American College President*. New York: McGraw-Hill.

Cohen, M., March, J., and Olsen, J. (1972) A garbage can model of organizational choice. *Administrative Science Quarterly* 17:1–25.

Community Energy Project. (1981) Community energy projects: how to plan them; how to run them. Fairfax, Va.: United Way of America.

Community Services Administration. (1980) *Too Cold . . . Too Dark: Rising Energy Prices and Low Income Households*. CSA Pamphlet 6143–18. Washington, D.C.: Author.

Corey, E. R. (1978) The Organizational Context of Industrial Buying Behavior. Working Paper Report No. 78–106. Cambridge, Mass.: Marketing Science Institute.

Cose, E. (1984) *Decentralizing Energy Decisions: The Rebirth of Community Power*. Boulder, Colo.: Westview.

Cottrell, F. (1955) *Energy and Society*. New York: McGraw-Hill.

Craig, C. S., and McCann, J. M. (1978) Assessing communication effects on energy conservation. *Journal of Consumer Research* 5:82–88.

Crecine, J. P. (1967) A computer simulation model of municipal budgeting. *Management Science* 13:786–815.

Cyert, R. M., and March, J. G. (1963) *A Behavioral Theory of the Firm*. Englewood Cliffs, N.J.: Prentice-Hall.

Cyert, R. M., Dill, W. R., and March, J. G. (1958) The role of expectations in business decision making. *Administrative Science Quarterly* 3:307–340.

Czepiel, J. A. (1974) Word-of-mouth processes in the diffusion of a major technological innovation. *Journal of Marketing Research* 11:172–180.

Daft, R. L., and Becker, S. W. (1978) *The Innovative Organization: Innovation Adoption in School Organizations*. New York: Elsevier North–Holland.

Darley, J. (1978) Energy conservation techniques as innovations and their diffusion. *Energy and Buildings* 1:339–343.

Darley, J. M., and Beniger, J. R. (1981) Diffusion of energy-conserving innovations. *Journal of Social Issues* 37(2):150–171.

Darley, J., and Berscheid, E. (1967) Increased liking as a result of the anticipation of personal contact. *Human Relations* 20:29–40.

Dearborn, D., and Simon, H. (1958) Selective perception: a note on the departmental identification of executives. *Sociometry* 21:140–144.

Department of Energy (1980) *Standby Gasoline Rationing Plan.* DOE/RG–0029. Washington, D.C.: Author.

Dickson, G. W. (1966) An analysis of vendor selection systems and decisions. *Journal of Purchasing* 2:5–17.

Dietz, T., and Vine, E. L. (1982) Energy impacts of a municipal conservation policy. *Energy* 7:755–758.

Dion, K. L. (1979) Intergroup conflict and intragroup cohesiveness. In W. G. Austin and S. Worchel, eds., *The Social Psychology of Intergroup Relations.* Monterey, Calif.: Brooks-Cole.

Dubin, R., and Spray, S. L. (1964) Executive behavior and interaction. *Industrial Relations* 3:99–108.

Dutton, W. H., and Kraemer, K. L. (1980) Automating bias. *Society* 17:36–41.

Dyer, L. M., and Fiske, S. T. (1982) Learning and schema development in organizations. Unpublished manuscript. Graduate School of Industrial Administration, Carnegie-Mellon University.

Dynes, R. R. (1970) *Organized Behavior on Disaster,* Lexington, Mass.: Heath.

Dynes, R. R. (1972) *A Perspective on Disaster Planning,* Columbus, Ohio: Disaster Research Center, Ohio State University.

Dynes, R. R., et al. (1979) Report of the Emergency Preparedness and Response Task Force, President's Commission on Three Mile Island, October.

Egan, J. R. (1982) To err is human factors. *Technology Review* February/March: 23–29.

Elgin, D., and Mitchell, A. (1977) Voluntary simplicity. *CoEvolution Quarterly* Summer: 5–18.

Ellsberg, D. (1972) *Papers on the War.* New York: Simon and Schuster.

Emerson, R. M. (1962) Power-dependence relations. *American Sociological Review* 27: 31–41.

Energy Information Administration. (1982) *Residential Energy Consumption Survey: Consumption and Expenditures April 1980 through March 1981. Part 1: National Data.* Washington, D.C.: U.S. Department of Energy, September (DOE/EIA–0321/1).

Engler, R. (1961) *The Politics of Oil: A Study of Private Power Democratic Directors.* New York: Macmillan.

Engler, R. (1977) *The Brotherhood of Oil: Energy Policy and the Public Interest.* New York: Mentor.

Ester, P., and Winett, R. A. (1982) Toward more effective antecedent strategies for environmental programs. *Journal of Environmental Systems* 11:201–221.

Farhar, B. C., Unseld, C. T., Vories, R., and Crews, R. (1980) Public opinion about energy. *Annual Review of Energy* 5:141–172.

Farhar, B. C., Weis, P., Unseld, C. T., and Burns, B. A. (1979) *Public Opinion About Energy: A Literature Review.* Golden, Colo.: Solar Energy Research Institute.

Federal Register. (1979) Residential Energy Conservation Program. 10 CFR, Part 456, Nov. 7, 1979, 44:64601.

Federal Register. (1981) Residential Energy Conservation Program. 10 CFR, Part 456, Nov. 12, 1981, 46:55836.

Federal Register. (1982) Residential Conservation Service Program. 10 CFR, Part 456, June 25, 1982, 47:27752.

Ferrey, S. E. (1981) Solar banking: constructing new solutions to the urban energy crisis. *Harvard Journal on Legislation* 18:483–551.

Festinger, L. (1957) *A Theory of Cognitive Dissonance.* Evanston, Ill.: Row, Peterson.

Festinger, L., and Aronson, E. (1960) Arousal and reduction of dissonance in social contexts. In D. Cartwright and A. Zander, eds., *Group Dynamics.* 2nd ed. New York: Harper and Row.

Finsterbusch, K., and Wolf, C. P. (1981) *Methodology of Social Impact Assessment.* Boston: Hutchinson Ross.

Fischer, G. W., and Crecine, J. P. (1980) How Much is Enough? Explaining Presidental Defense Budget Requests. Department of Social Science Working Paper, Carnegie-Mellon University.

Fischhoff, B. (1975a) Hindsight/foresight: the effect of outcome knowledge on judgment under uncertainty. *Journal of Experimental Psychology: Human Perception and Performance* 1:288–299.

Fischhoff, B. (1975b) Hindsight: thinking backward? *Psychology Today* 8:70–76.

Fischhoff, B., and Beyth, R. (1975) I knew it would happen: remembered probabilities of once-future things. *Organizational Behavior and Human Performance* 13:1–16.

Fitchburg Office of the Planning Coordinator. (1980) Fitchburg Action to Conserve Energy (FACE), Final Report. Fitchburg, Mass.: Author.

Forrester, J. (1961) *Industrial Dynamics.* Cambridge, Mass.: MIT Press.

Freedman, J., and Fraser, S. (1966) Compliance without pressure: the foot-in-the-door technique. *Journal of Personality and Social Psychology* 4:195–202.

Freudenberg, W. R. (1982) The effects of rapid population growth on the social and personal well-being of local community residents. In B. A. Weber and R. E. Howell, eds., *Coping with Growth in Rural Communities.* Boulder, Colo: Westview.

Friedman, J. J. (1977) Community action on water pollution. *Human Ecology* 5:329–353.

Fritz, C. E. (1961) Disaster. In R. K. Merton and R. A. Nisbet, eds., *Contemporary Social Problems.* 651–694. New York: Harcourt.

Fritz, C. E. (1967) Disasters. In D. L. Sills, eds., *International Encyclopedia of the Social Sciences.* Vol. 4:202–207. New York: Macmillan and Free Press.

General Accounting Office. (1981) *Residential Energy Conservation Outreach Activities—A New Federal Approach Needed.* Washington, D.C.: Author, February 11.

Gerard, H., and Mathewson, G. (1966) The effects of severity of initiation on liking for a group: a replication. *Journal of Experimental Social Psychology* 2:278–287.

Goldman, M., Stockbauer, J. W., and McAuliffe, T. G. (1977) Intergroup and intragroup cooperation and competition. *Journal of Experimental Social Psychology* 13: 81–88.

Haber, R. N., ed. (1969) *Information-Processing Approaches to Visual Perception.* New York: Holt.

Hall, D., and Mansfield, R. (1971) Organizational and individual response to external stress. *Administrative Science Quarterly* 16:533–547.

Hall, R. I. (1976) A system pathology of an organization: the rise and fall of the old *Saturday Evening Post. Administrative Science Quarterly* 21:185–211.

Halperin, M. H. (1974) *Bureaucratic Politics and Foreign Policy.* Washington, D.C.: Brookings Institution.

Hamill, R., Wilson, T. D., and Nisbett, R. (1980) Insensitivity of sample bias: generalizing from a typical case. *Journal of Personality and Social Psychology* 39(4):578–589.

Hayes, D. (1976) *The Case for Conservation.* Worldwatch Paper No. 2. Washington, D.C.: Worldwatch Institute.

Hayward, G. (1972) Diffusion of innovation in the flour milling industry. *European Journal of Marketing* 6:195–202.

Heberlein, T. A., and Warriner, G. K. (1982) The Influence of Price and Attitude on Shifting Residential Electricity Consumption from On to Off-peak Periods. Paper presented at the International Conference on Consumer Behaviour and Energy Policy, Noordwijkerhout, Netherlands, September.

Hershey, R. D., Jr. (1981) Maneuvers on energy unit. *New York Times*, September 14, p. IV–2.

Hershey, R. D., Jr. (1982) The dark side of the oil glut. *New York Times*, March 21.

Higbee, K. L. (1969) Fifteen years of fear arousal: research on threat appeals: 1953–1968. *Psychological Bulletin* 72:426–444.

Hirst, E. (1976) Transportation energy conservation policies. *Science* 192:15–20.

Hirst, E., Berry, L., and Soderstrom, J. (1981) Review of utility home energy audit programs. *Energy* 6:621–630.

Hoos, I. (1978) The credibility issue. In *Essays on Issues Relevant to the Regulation of Radioactive Waste Management*. Nuclear Regulatory Commission, NUREG–0412, April.

Hovland, C., and Weiss, W. (1951) The influence of source credibility on communication effectiveness. *Public Opinion Quarterly* 15:635–650.

Hovland, C. I., Janis, I. L., and Kelley, H. H. (1953) *Communication and Persuasion*. New Haven, Conn.: Yale University Press.

Hsu, S. (1979) The Diffusion of Energy-Efficient Technologies in Industry. Report to U. S. Department of Energy. State University of New York, Binghamton.

Hutton, R. B. (1982) Advertising and the Department of Energy's campaign for energy conservation. *Journal of Advertising* 11(2):27–39.

Irle, M., and Montmann, V., eds. (1978) *Theorie der Kognitiven Dissonanz* (by L. Festinger, originally published 1957). Bern: Huber.

Jacobs, D. (1974) Dependence and vulnerability: an exchange approach to the control of organizations. *Administrative Science Quarterly* 19:45–59.

Johnson, H. T. (1978) Management accounting in an early multidivisional organization: General Motors in the 1920s. *Business History Review* 52:490–517.

Joint Economic Committee. (1977) Testimony of Lester Thurow, in *Hearings on the Economics of the President's Proposed Energy Policies Before the Joint Economic Committee*, 95th Congress, 1st Session, 22.

Jones, E., and Kohler, R. (1958) The effects of plausibility on the learning of controversial statements. *Journal of Abnormal and Social Psychology* 57:315–320.

Jones, E. E., and Nisbett, R. E. (1971) The actor and the observer: divergent perceptions of the causes of behavior. In E. E. Jones, D. Kanouse, H. H. Kelley, R. E. Nisbett, S. Valins, and B. Weiner, eds., *Attribution: Perceiving the Causes of Behavior*. Morristown, N.J.: General Learning Press.

Kahneman, D., Slovic, P., and Tversky, A., eds. (1982) *Judgment Under Uncertainty: Heuristics and Biases*. Cambridge: Cambridge University Press.

Kamlet, M. S., and Mowery, D. C. (1980) The budgetary base in federal resource allocation. *American Journal of Political Science* 24:804–821.

Kash, D. E., et al. (1976) *Our Energy Future*. Norman: University of Oklahoma Press.

Katz, E. (1961) The social itinerary of technical change: two studies on the diffusion of innovation. *Human Organization* 20:70–82.

Katz, E., and Lazarsfeld, P. F. (1955) *Personal Influence*. Chicago: The Free Press of Glencoe, Ill.

Katzev, R. D., and Johnson, T. R. (1982) A Social Psychological Analysis of Residential Electricity Consumption: The Impact of Minimal Justification Techniques. Paper presented at the International Conference on Consumer Behaviour and Energy Policy, Noordwijkerhout, Netherlands, September.

Kaufman, H. (1971) *The Limits of Organizational Change*. University, Ala.: University of Alabama Press.

Kaufman, H. (1977) Reflections on administrative reorganization. In J. A. Pechman, ed., *Setting National Priorities: The 1978 Budget*. Washington, D.C.: Brookings Institution.

Kay, N. M. (1979) *The Innovating Firm: A Behavioral Theory of Corporate R&D*. New York: St. Martin's Press.

Keegan, W. (1974) Multinational scanning: a study of information sources utilized by head-quarters executives in multinational companies. *Administrative Science Quarterly* 19:411–421.

Kempton, W., and Montgomery, L. (1982) Folk quantification of energy. *Energy* 7:817–827.

Kempton, W., Harris, C. K., Keith, J. G., and Weihl, J. S. (1982) Do Consumers Know "What Works" in Energy Conservation? Paper presented at the American Council for an Energy-Efficient Economy Summer Study, Santa Cruz, California, August.

Kiesler, S., and Sproull, L. (1982) Managerial response to changing environments: perspectives on problem sensing from social cognition. *Administrative Science Quarterly* 27:548–570.

Kiser, G. E., and Rao, C. P. (1977) Important vendor factors in industrial and hospital organizations: a comparison. *Industrial Marketing Management* 6:289–296.

Knox, R., and Inkster, J. (1968) Postdecision dissonance at post time. *Journal of Personality and Social Psychology* 8:319–323.

Koster, F. (1981) An analysis of mass media campaigns designed to educate the public about energy use. Jacksonville, Fla.: Koster/Hopkins, Inc.

Kotler, P. (1975) *Marketing for Nonprofit Organizations.* Englewood Cliffs, N.J.: Prentice-Hall.

Kotler, P., and Zaltman, G. (1971) Social marketing: an approach to planned social change. *Journal of Marketing* 35(July):3–12.

Kreps, G. (1978) The social organization of disaster response. In E. L. Quarantelli, ed., *Disasters: Theory and Research.* London: Eege.

Kreps, G. (1979) The Worth of the NAS-NRC (1952–63) and DRC (1963–present) Studies of Individual and Social Response to Disasters. Paper presented at the Social and Demographic Research Institute Conference on Disasters, National Science Foundation, Washington, D.C., May 18–19.

Kulp, G., Shonka, D. B., Collins, M. J., Murphy, B. J., and Reed, K. J. (1980) *Transportation Energy Conservation Data Book: Edition 4.* Oak Ridge, Tenn.: Oak Ridge National Laboratory (ORNL–5654), September.

Kunreuther, H. (1979) Changing societal consequences of risks from natural hazards, *Annals of the American Academy of Political and Social Science* 443:104–116.

Kunreuther, H., et al. (1978) *Disaster Insurance Protection: Public Policy Lessons.* New York: Wiley.

Lane, R. E. (1953) Why businessman violate the law. *Journal of Criminal Law, Criminology and Police Science* 44:151–165.

Langer, E. J., and Rodin, J. (1976) The effects of choice and enhanced responsibility for the aged: a field experiment in an institutional setting. *Journal of Personality and Social Psychology* 34:191–198.

Larkey, P. D. (1979) *Evaluating Public Programs: The Impact of General Revenue Sharing on Municipal Government.* Princeton, N.J.: Princeton University Press.

Larkey, P. D., and Smith, R. A. (1981) Formulating and Justifying Budget Problems: Bad News and Not-So-Good News. Prepared for the First Annual Symposium on Information Processing in Organizations, Carnegie-Mellon University, Pittsburgh, Pennsylvania.

Latané, B., and Darley, J. M. (1968) Group inhibition of bystander intervention in emergencies. *Journal of Personality and Social Psychology* 10:215–221.

Latané, B., and Darley, J. M. (1970) *The Unresponsive Bystander: Why Doesn't He Help?* New York: Appleton-Century-Crofts.

Latané, B., and Nida, S. (1981) Ten years of research on group size and helping. *Psychological Bulletin* 89:308–324.

Lee, D., Jr. (1973) Requiem for large-scale models. *Journal of the American Institute of Planners* 39:136–178.

Lehmann, D. R., and O'Shaughnessy, J. (1974) Difference in attribute importance for different industrial products. *Journal of Marketing* 38:36–42.

Leonard-Barton, D. (1980) The Role of Interpersonal Communication Networks in the

Diffusion of Energy Conserving Practices and Technologies. Unpublished paper, Sloan School of Management, Massachusetts Institute of Technology.

Leonard-Barton, D. (1981a) The diffusion of active residential solar energy equipment in California. In A. Shama, ed., *Marketing Solar Energy Innovations.* New York: Praeger.

Leonard-Barton, D. (1981b) Voluntary simplicity lifestyles and energy conservation. *Journal of Consumer Research* 8:243–252.

Leonard-Barton, D., and Rogers, E. M. (1979) Adoption of Energy Conservation Among California Homeowners. Paper presented to the International Communication Association, Philadelphia.

Leonard-Barton, D., and Rogers, E. M. (1981) Horizontal Diffusion of Innovations: An Alternative Paradigm to the Classical Diffusion Model. Working Paper 1214–81, Sloan School of Management, Massachusetts Institute of Technology.

Leventhal, H. (1970) Findings and theory in the study of fear communications. In L. Berkowitz, ed., *Advances in Experimental Social Psychology.* Vol. 5. New York: Academic Press.

Lewis, W. (1980) *Reducing U.S. Oil Vulnerability: Energy Policy for the 1980's.* Washington, D.C.: U.S. Department of Energy.

Lindberg, L. (1977) *The Energy Syndrome.* Lexington, Mass.: Lexington Books.

Lindblom, C. E. (1959) The science of muddling through. *Public Administration Review* 19:79–88.

Lovins, A. (1977) *Soft Energy Paths: Toward a Durable Peace.* Cambridge, Mass.: Ballinger.

Lovins, A. B., and Lovins, L. H. (1981) Economically Efficient Solar Futures. Paper presented at the meeting of the American Association for the Advancement of Science, Toronto, January.

Lowe, A. E., and Shaw, R. W. (1968) An analysis of managerial biasing: evidence from a company's budgeting process. *The Journal of Management Studies* 5:304–315.

Lowenthal, M. M. (1981) Roosevelt and the coming of the war: the search for United States policy 1937–1942. *Journal of Contemporary History* 16:413–440.

Lowi, T. J. (1964) American business, public policy, case-studies, and political theory. *World Politics* 16(4):677–715.

Lowi, T. J. (1972) Four systems of policy, politics, and choice. *Public Administration Review* 32(4):298–310.

Lundstrom, E. (1980) Energy consumption in single-family houses—influence of the occupants. Royal Institute of Technology, Department of Building Economics and Organization, Stockholm, Sweden.

Lyles, M. A., and Mitroff, I. I. (1980) Organizational problem formulation: an empirical study. *Administrative Science Quarterly* 25:102–119.

Manns, C. L., and March, J. G. (1978) Financial adversity, internal competition and curriculum change in a university. *Administrative Science Quarterly* 23:541–552.

Mansfield, E. (1961) Technical change and the rate of innovation. *Econometrica* 29:741–766.

Mansfield, E. (1963) Size of firm, market structure and innovation. *Journal of Political Economy* 71:566–576.

Mansfield, E. (1964) Industrial research and development expenditures: determinants, prospects and relation to size of firm and inventive output. *Journal of Political Economy* 72:319–340.

Mansfield, E. (1968) *Industrial Research and Technological Innovation: An Econometric Analysis.* New York: Norton.

March, J. G. (1981) Decisions in organizations and theories of choice. In A. Van de Ven and W. Joyce, eds., *Assessing Organizational Design and Performance.* New York: Wiley Interscience.

March, J. G., and Olsen, J. P. (1976) *Ambiguity and Choice in Organizations.* Bergen, Norway: Universitetsforlaget.

March, J. G., and Olsen, J. P. (1982) Organizing Political Life: What Administrative Reorganization Tells us About Governing. Paper presented at the meeting of the American Political Science Association, Denver, Colorado, September.

March, J. G., and Shapira, Z. (1982) Behavioral decision theory and organizational decision theory. In G. R. Ungson and D. N. Braunstein, eds., *Decision Making: An Interdisciplinary Inquiry*. Boston: Kent Publishing Company.

March, J. G., and Simon, H. A. (1958) *Organizations*. New York: Wiley.

Martin, D. (1982) Energy shortage eases materially; basic shifts in consumption cited. *New York Times*, March 8, pp. A1, D20.

Marley, R. (1982) Energy Trends in the U.S. Economy. Office of Policy, Planning, and Analysis, U.S. Department of Energy.

Mazis, M. R. (1975) Antipollution measures and psychological reactance theory: a field experiment. *Journal of Personality and Social Psychology* 31:654–660.

McClelland, L. (1980) Encouraging Energy Conservation in Multifamily Housing: RUBS and Other Methods of Allocating Energy Costs to Residents. Executive Summary and List of Contents. Prepared for the U.S. Department of Energy. Boulder, Colo.: Institute of Behavioral Science, University of Colorado.

McClelland, L. (1982) Is Energy Efficiency a Factor in the Rental Housing Market? Paper presented to the American Association for the Advancement of Science, Washington, D.C.

McGuire, W. J. (1969) The nature of attitudes and attitude change. In G. Lindzey and E. Aronson, eds., *The Handbook of Social Psychology*. 2nd ed., Vol. 3. Reading, Mass.: Addison-Wesley.

McGuire, W. J. (1983) Attitudes and attitude change. In G. Lindzey and E. Aronson, eds., *Handbook of Social Psychology*. 3rd ed. Reading, Mass.: Addison-Wesley.

McNutt, B., and Rucker, E. (1981) Impact of Fuel Economy Information on New Car and Light Truck Buyers. U.S. Department of Energy, Office of Conservation Policy Planning and Analysis, and Office of Transportation Programs, Conservation and Renewable Energy.

Meade, J. E. (1970) *Theory of Indicative Planning*. Manchester: Manchester University Press.

Mettler-Meibom, B., and Wichmann, B. (1982) The influence of information and attitudes toward energy conservation on behavior. Translated by M. Stommel. In H. Schaefer, ed., *Einfluss des Verbraucherverhaltens auf den Energiebedarf Privater Haushalte*. Berlin: Springer-Verlag.

Mileti, D. S. (1975) *Natural Hazard Warning Systems in the United States: A Research Assessment*. Boulder: Institute of Behavioral Science, University of Colorado.

Mills, M. K. (1981) Energy Conservation and the Retail Industry: Public Policy Implications. Unpublished paper, Department of Marketing, University of Southern California.

Milstein, J. S. (1978) Soft and Hard Energy Paths: What People on the Streets Think. Unpublished report, Office of Conservation and Solar Applications, U.S. Department of Energy, March.

Mintzberg, H. (1973) *The Nature of Managerial Work*. New York: Harper and Row.

Mintzberg, H. (1978) Patterns in strategy formation. *Management Science* 24:934–948.

Moch, M. K., and Morse, E. V. (1977) Size, centralization and the adoption of innovations. *American Sociological Review* 42:716–725.

Mowery, D. C., Kamlet, M. S., and Crecine, J. P. (1980) Presidential management of budgetary and fiscal policymaking. *Political Science Quarterly* 95:395–425.

Murray, J. R., Minor, M. J., Bradburn, N. M., Cotterman, R. F., Frankel, M., and Pisarski, A. E. (1974) Evaluation of public response to the energy crisis. *Science* 184:257–263.

Mynatt, F. R. (1982) Nuclear reactor safety research since Three Mile Island. *Science* 216: 131–136.

Nader, L., and Beckerman, S. (1978) Energy as it relates to the quality and style of life. *Annual Review of Energy* 3:1–28.

Nader, L., and Milleron, N. (1979) Dimension of the "people problem" in energy research and "the" factual basis of dispersed energy futures. *Energy* 4:953–967.

National Academy of Sciences. (1981) *Energy Efficiency: The Impact of Conservation. An Academy Forum, February 4, 1981.* Washington, D.C.: National Academy Press.

National Research Council. (1979) *Energy in Transition 1985–2010.* Washington, D.C.: Author.

National Research Council. (1980) *Energy Choices in a Democratic Society.* Supporting Paper 7, Report of the Consumption, Location, and Occupational Patterns Resource Group, Synthesis Panel of the Committee on Nuclear and Alternative Energy Systems. Washington, D.C.: National Academy of Sciences.

Neels, K. (1981) Effects of Alternate Metering Arrangements on Energy Use and Upgrading of Rental Property: Progress Report. Unpublished draft. Santa Monica, Calif.: Rand Corporation.

Neels, K. (1982) Reducing energy consumption in housing: An assessment of alternatives. *International Regional Science Review* 7:69–82.

Nelkin, D. (1981) Nuclear power as a feminist issue. *Environment* 23(1):14–20, 38–39.

Nelkin, D., and Fallows, S. (1978) The evolution of the nuclear debate: the role of public participation. *Annual Review of Energy* 3:275–312.

Neustadt, R. E., and Fineberg, H. V. (1978) The Swine Flu Affair: Decision Making on a Slippery Disease. Washington, D.C.: U.S. Department of Health, Education, and Welfare.

New York Times (1981) March 11, p. II:4.

Newtson, D. (1973) Attribution and the unit of perception of ongoing behavior. *Journal of Personality and Social Psychology* 28:28–38.

Newtson, D., and Rinder, R. (1979) Variation in behavior perception and ability attribution. *Journal of Personality and Social Psychology* 37:379–388.

Nisbett, R. E., Borgida, E., Crandall, R., and Reed, H. (1976) Popular induction: information is not necessarily informative. In J. S. Carroll and J. W. Payne, eds., *Cognition and Social Behavior.* Hillsdale, N.J.: Lawrence Erlbaum Associates.

Office of Policy, Planning, and Analysis. (1982) Briefing materials on trends in energy use and conservation. Washington, D.C.: U.S. Department of Energy, April 14.

Office of Program Coordination (1981) *Directory of Social Science Expertise in the Department of Energy: A Guide to Agency Staff Trained or Working in the Social Sciences.* Office of Environment, U.S. Department of Energy. Washington, D.C.: Author.

Office of Technology Assessment (1980) *Residential Energy Conservation.* Montclair, N.J.: Allanheld, Osmun.

Office of Technology Assessment (1982) *Energy Efficiency of Buildings in Cities.* Washington, D.C.: Author.

Olsen, M. E. (1981) Consumers' attitudes toward energy conservation. *Journal of Social Issues* 37(2):108–131.

Olsen, M., and Cluett, C. (1979) Evaluation of the Seattle City Light Neighborhood Energy Conservation Program. Seattle, Wash.: Battelle Human Affairs Research Center.

O'Reilly, C. A., and Weitz, B. (1980) Managing marginal employees: the use of warnings and dismissals. *Administrative Science Quarterly* 25:467–484.

OR/MS Dialog (1980) A Market Assessment for Active Solar Heating and Cooling Products. Cambridge, Mass.: Author.

Oskamp, S. (1981) Energy conservation by industrial and commercial users: two surveys. *Journal of Environmental Systems* 10:201–213.

Ozanne, U. B., and Churchill, G. A. (1978) Adoption research: information sources in the industrial purchasing decision. In R. L. King, ed., *Marketing and the New Science of Planning.* Chicago: American Marketing Association.

Padgett, J. F. (1980a) Bounded rationality in budgetary research. *American Political Science Review* 74:354–372.

Padgett, J. F. (1980b) Managing garbage can hierarchies. *Administrative Science Quarterly* 25:583–604.

Padgett, J. F. (1981) Hierarchy and ecological control in federal budgetary decision making. *American Journal of Sociology* 87:75–129.

Pallak, M. S., Cook, D. A., and Sullivan, J. J. (1980) Commitment and energy conservation. In L. Bickman, ed., *Applied Social Psychology Annual.* Vol. 1. Beverly Hills: Sage Publications.

Palmer, D. A. (1980) Broken Ties: Some Political and Interorganizational Determinants of Interlocking Directorates among Large American Corporations. Paper presented at the annual meeting of the American Sociological Association, New York.

Penz, A. J. (1981) Searching for Home Energy Improvements: A Household's Perspective. Pittsburgh, Pa.: Institute of Building Sciences, Carnegie-Mellon University.

Perrow, C. (1981) Normal accident at Three Mile Island. *Society* 18:17–26.

Pettigrew, A. M. (1972) Information as power and control. *Sociology* 6:187–204.

Pfeffer, J. (1972a) Interorganizational influence and managerial attitudes. *Academy of Management Journal* 15:317–330.

Pfeffer, J. (1972b) Merger as a response to organizational interdependence. *Administrative Science Quarterly* 17:382–394.

Pfeffer, J. (1978) *Organizational Design.* Arlington Heights, Ill.: AHM Publishing.

Pfeffer, J., and Leblebici, H. (1973) The effect of competition on some dimensions of organizational structure. *Social Forces* 52:268–279.

Pfeffer, J., and Nowak, P. (1976) Joint ventures and interorganizational dependence. *Administrative Science Quarterly* 21:398–418.

Pfeffer, J., and Salancik, G. R. (1978) *The External Control of Organizations: A Resource Dependence Perspective.* New York: Harper & Row.

Pierce, J. L., and Delbecq, A. L. (1977) Organization structure, individual attitudes and innovation. *Academy of Management Review* 2:27–37.

Pirkey, D. (1982) Assessment of the Federal Fuel Economy Information Program. Washington, D.C.: U.S. Department of Energy, Office of Policy, Planning and Analysis.

Pirkey, D., McNutt, B., Hemphill, J., and Dulla, R. (1982) Consumer Response to Fuel Economy Information—Alternative Sources, Uses and Formats. Paper presented to Society of Automotive Engineers, Detroit.

Pounds, W. (1969) The process of problem finding. *Industrial Management Review* 11:1–19.

President's Commission on the Accident at Three Mile Island (1979) *The Need for Change: The Legacy of TMI.* Washington, D.C.: Author.

Quarantelli, E., and R. R. Dynes (1977) Response to social crisis and disaster. *Annual Review of Sociology* 3:23–49.

Quelch, J. A., and Thirkell, P. (1978) "Imposed Choice" Purchases of Energy-Using Equipment in the Residential Sector: A Review. Unpublished paper, School of Business Administration, The University of Western Ontario.

Quinney, R. (1970) *The Social Reality of Crime.* Boston: Little, Brown.

Rankin, W. L. (1978) Human Value Conflicts Concerning Nuclear Power. Paper presented at the meeting of the American Psychological Association, Toronto, August.

Reich, J. W., and Robertson, J. L. (1979) Reactance and norm appeal in anti-littering messages. *Journal of Applied Social Psychology* 9:91–101.

Reichel, D. A., and Geller, E. S. (1981) Applications of behavioral analysis for conserving transportation energy. In A. Baum and J. E. Singer, eds., *Advances in Environmental Psychology.* Vol. 3. *Energy Conservation: Psychological Perspectives.* Hillsdale, N.J.: Lawrence Erlbaum Associates.

Ritchey, F. J. (1981) Medical rationalization, cultural lag, and the malpractice crisis. *Human Organizations* 40:97–112.

Robinson, J. B. (1982) Apples and horned toads: on the framework-determined nature of the energy debate. *Policy Sciences* 15:23–45.

Rochlin, G. I. (1977) Nuclear waste disposal: two social criteria. *Science* 195:23–30.

Rodberg, L., and Schachter, M. (1980) *State Conservation and Solar Energy Tax Programs: Incentives or Windfalls?* Washington, D.C.: Council of State Planning Agencies.

Rodin, J., and Langer, E. J. (1977) Long-term effects of a control-relevant intervention with the institutionalized aged. *Journal of Personality and Social Psychology* 35: 897–902.

Rohr, J. A. (1981) Financial disclosure: power in search of policy. *Public Personnel Management Journal* 40:29–40.

Rogers, E. M., with Shoemaker, F. (1971) *The Communication of Innovations.* New York: Free Press.

Roscoe, S. N., ed. (1980) *Aviation Psychology.* Ames: Iowa State University Press.

Rosenbaum, W. A. (1981) *Energy Politics and Public Policy.* Washington, D.C.: Congressional Quarterly Press.

Rosenberg, M. (1980) The RCS Project—Preliminary Findings and Recommendations. Boston: Technical Development Corporation.

Ross, L. (1977) The intuitive psychologist and his shortcomings: distortions in the attribution process. In L. Berkowitz, ed., *Advances in Experimental Social Psychology.* Vol. 10. New York: Academic Press.

Ross, M. H., and Williams, R. H. (1981) *Our Energy: Regaining Control.* New York: McGraw-Hill.

Russo, J. E. (1977) A proposal to increase energy conservation through provision of consumption and cost information to consumers. In B. A. Greenberg and D. N. Bellenger, eds., *Contemporary Marketing Thought: 1977 Educators' Proceedings.* Chicago: American Marketers Association.

Salancik, G. R. (1979) Interorganizational dependence and responsiveness to affirmative action: the case of women and defense contractors. *Academy of Management Journal* 22:375–394.

Sant, R. W. (1979) *The Least-Cost Energy Strategy.* Pittsburgh: Carnegie-Mellon University Press.

Sant, R. W., Carhart, S. C., with Bakke, D. W., and Mulherkar, S. S. (1981) *Eight Great Energy Myths: The Least-Cost Energy Strategy—1978–2000.* Energy Productivity Report No. 4, The Energy Productivity Center. Arlington, Va.: Mellon Institute.

Schiff, A. L. (1966) Innovation and administrative decision making: the conservation of land resources. *Administrative Science Quarterly* 11:1–30.

Schiff, M. (1970) The impact of people of budgets. *The Accounting Review* 45:259–268.

Schiff, M., and Lewin, A. Y. (1968) Where traditional budgeting fails. *Financial Executive* 36:51–62.

Schipper, L. (1976) Rising the productivity of energy utilization. *Annual Review of Energy* 1:455–517.

Schipper, L., and Ketoff, A. (1982) Energy Conservation in the OECD. Paper presented at the American Council for an Energy-Efficient Economy Summer Study, Santa Cruz, California, August.

Schnaiberg, A. (1982) Energy Conservation in the U.S.: A Pyrrhic Social Victory? Unpublished manuscript, Department of Sociology, Northwestern University, January.

Schott, B., and von Grebner, K. (1974) R&D, innovation and microeconomic growth: a case study. *Research Policy* 2:380–403.

Schultz, D. P. (1965) Theories of panic behavior: a review. *Journal of Social Psychology* 66:31–40.

Schwartz, S. H., and Clausen, G. T. (1970) Responsibility, norms, and helping in an emergency. *Journal of Personality and Social Psychology* 16:299–310.

Science Applications, Inc. (1982) Consumer Products Efficiency Standard Economic Analysis Document. Washington, D.C.: U.S. Department of Energy.

Seligman, C., and Becker, L., eds. (1981) Energy conservation. *Journal of Social Issues* 37(2).

Seligman, C., Becker, L. J., and Darley, J. M. (1981) Encouraging residential energy conservation through feedback. In A. Baum and J. E. Singer, eds., *Advances in Environmental Psychology.* Vol. 3. *Energy Conservation: Psychological Perspectives.* Hillsdale, N.J.: Lawrence Erlbaum Associates.

Sheridan, T. B., and Johannsen, G., eds. (1976) *Monitoring Behavior and Supervisory Control.* New York: Plenum.

Sherif, M. (1935) A study of some social factors in perception. *Archives of Psychology* 187.

Sherif, M. Harvey, O. J., White, B. J., Hood, W. R., and Sherif, C. W. (1961) *Intergroup Conflict and Cooperation: The Robbers Cave Experiment.* Norman, Okla.: Institute of Group Relations, University of Oklahoma.

Shippee, G. (1980) Energy consumption and conservation psychology: a review and conceptual analysis. *Environmental Management* 4:297–314.

Simon, H. A. (1957) *Models of Man: Social and Rational.* New York: Wiley.

Simon, H. A. (1971) Designing organizations for an information-rich world. In M. Greenberger, ed., *Computers, Communications, and the Public Interest.* Baltimore: Johns Hopkins Press.

Simon, H. A. (1973) Applying information technology to organization design. *Public Administration Review* 33:268–278.

Slovic, P., Fischhoff, B., and Lichtenstein, S. (1978) Accident probabilities and seat belt usage: a psychological perspective. *Accident Analysis and Prevention* 10:281–285.

Slovic, P., Fischoff, B. and Lichtenstein, S. (1982) Facts versus fears: understanding perceived risks. In D. Kahneman, P. Slovic, and A. Tversky, eds., *Judgment Under Uncertainty: Heuristics and Biases.* Cambridge: Cambridge University Press.

Slovic, P., Fischoff, B., Lichtenstein, S., Corrigan, B., and Combs, B. (1977) Preference for insuring against probable small losses: insurance implications. *Journal of Risk and Insurance* 44:237–258.

Slovic, P., Kunreuther, H., and White, G. (1974) Decision processes, rationality, and adjustment to natural hazards. In G. White, ed., *Natural Hazards: Local, National, Global.* New York: Oxford.

Smith, R. J. (1982) Edwards defends budget cuts at DOE. *Science,* 216:716–717.

Solar Energy Research Institute (1981) *A New Prosperity: Building a Sustainable Energy Future.* Andover, Mass.: Brick House.

Sonderegger, R. C. (1978) Movers and stayers: the resident's contribution to variation across houses in energy consumption for space heating. In R. H. Socolow, ed., *Saving Energy in the Home: Princeton's Experiments at Twin Rivers.* Cambridge, Mass.: Ballinger.

Spector, M., and Kitsuse, J. I. (1977) *Constructing Social Problems.* Menlo Park, Calif.: Cummings.

Sproull, L. S. (1981a) The Nature of Managerial Attention. Prepared for the First Annual Symposium on Information Processing in Organizations, Carnegie-Mellon University.

Sproull, L. S. (1981b) Managing educational programs: a micro-behavioral analyis. *Human Organizations* 40:113–122.

Sproull, L., Weiner, S., and Wolf, D. (1978) *Organizing an Anarchy.* Chicago: University of Chicago Press.

Stanton, T. (1982) Lessons and accomplishments of community-wide volunteer energy conservation campaigns. Paper presented at American Council for an Energy Efficient Economy Summer Study, Santa Cruz, Calif., August.

Starbuck, W. H., Arent, G., and Hedberg, B. L. T. (1978) Responding to crises. *Journal of Business Administration* 9:111–137.

Staub, E. (1974) Helping a distressed person: social, personality, and stimulus determinants. In L. Berkowitz, ed., *Advances in Experimental Social Psychology.* Vol. 7. New York: Academic Press.

Staw, B. M., and Szwajkowski, E. (1975) The scarcity-munificence component of organizational environments and the commission of illegal acts. *Administrative Science Quarterly* 20:345–354.

Staw, B. M., Sandelands, L. E., and Dutton, J. E. (1981) Threat of rigidity effects in organizational behavior: a multilevel analysis. *Administrative Science Quarterly* 26: 501–524.

Stern, L. W., and Morgenroth, W. M. (1968) Concentration, mutually recognized interdependence, and the allocation of marketing resources. *Journal of Business* 41:56–67.

Stern, P. C., and Gardner, G. T. (1981) Psychological research and energy policy. *American Psychologist* 36(4):329–342.

Stern, P. C., and Kirkpatrick, E. M. (1977) Energy behavior. *Environment* 19(9):10–15.

Stern, P. C., Black, J. S. and Elworth, J. T. (1981) *Home Energy Conservation: Issues and Programs for the 1980s.* Mount Vernon, N.Y.: Consumers Union Foundation.

Stern, P. C., Black, J. S., and Elworth, J. T. (1982a) Influences on Household Energy Adaptations. Paper presented to the American Association for the Advancement of Science, Washington, D.C.

Stern, P. C., Black, J. S., and Elworth, J. T. (1982b) Personal and Contextual Influences on Household Energy Adaptations. Paper presented at the International Conference on Consumer Behavior and Energy Policy, Noordwijkerhout, Netherlands, September.

Stern, P. C., Black, J. S., and Elworth, J. T. (1983) Responses to changing energy conditions among Massachusetts households. *Energy* 8:515–523.

Stobaugh, R., and Yergin, D., eds. (1979) *Energy Future.* New York: Random House.

Stolwijk, J. A. J., ed. (1978) *Energy Conservation Strategies in Buildings.* New Haven: John B. Pierce Foundation of Connecticut, Inc.

Sutton, F. X. (1969) Technical assistance. In D. Sills, ed., *International Encyclopedia of the Social Sciences.* Vol. 15. 565–576. New York: Macmillan and Free Press.

Talbot, D., and Morgan, R. E. (1981) *Power and Light: Political Strategies for the Solar Transition.* New York: Pilgrim Press.

Taylor, S. E., and Fiske, S. T. (1978) Salience, attention, and attribution: top of the head phenomena. In L. Berkowitz, ed., *Advances in Experimental Social Psychology.* Vol. 11. New York: Academic Press.

Taylor, S. E., and Thompson, S. C. (1982) Stalking the elusive "vividness" phenomenon. *Psychological Review* 89:155–181.

Thomas, R., Petter, D., Spurway, D., and Etzler, R. (1975) Gasaver Evaluation. Falls Church, Va.: U.S. Army Operational Test and Evaluation Agency.

Thompson, J. D. (1967) *Organizations in Action.* New York: McGraw-Hill.

Thompson, S. (1981) Will it hurt less if I can control it? A complex answer to a simple question. *Psychological Bulletin* 90:89–101.

Trippi, R. R., and Wilson, D. R. (1974) Technology transfer and the innovative process in small entrepreneurial organizations. *Journal of Economics and Business* 27:64–68.

Turner, B. A. (1976) The organizational and interorganizational development of disasters. *Administrative Science Quarterly* 21:378–397.

Turner, R. H. (1980) The mass media and preparation for natural disaster. In *Disasters and the Mass Media.* Washington, D.C.: National Academy of Sciences.

Tushman, M., and Scanlon, T. J. (1981) Characteristics and external orientations of boundary spanning individuals. *Academy of Management Journal* 24:83–98.

Tversky, A., and Kahneman, D. (1974) Judgment under uncertainty: heuristics and biases. *Science* 185:1124–1131.

Tversky, A., and Kahneman, D. (1981) The framing of decisions and the psychology of choice. *Science* 211:453–458.

U. S. Department of Transportation. (1975) A History of the Energy Crisis: October 1973 through March 1974. Transportation Systems Center Working Papers, Cambridge, Massachusetts, May.

Unseld, C. T., Morrison, D. E., Sills, D. L., and Wolf, C. P., eds. (1979) *Scoiopolitical Effects of Energy Use and Policy.* Supporting Paper 5. Reports to the Sociopolitical Effects Resource Group, Risk and Impact Panel of the Committee on Nuclear and Alternative Energy Systems. Washington, D.C.: National Academy of Sciences.

Van Cott, H. P., and Kinkade, R. G., eds. (1972) *Human Engineering Guide to Equipment Design.* Washington: American Institutes for Research.

Verhallen, T. M. M., and van Raaij, W. F. (1981) Household behavior and the use of natural gas for home heating. *Journal of Consumer Research* 8:253–257.

Walker, J. M. (1980) Voluntary response to energy conservation appeals. *Journal of Consumer Research* 7:88–92.

Warkov, S. (1978) *Energy Policy in the United States: Social and Behavioral Dimensions.* New York: Praeger.

Watney, J. (1974) *Clive in India.* England: Saxon House.

Webster, F. E. (1970) Informal communication in industrial markets. *Journal of Marketing Research* 7 (May):186–189.

Webster, F. E. (1971) Communication and diffusion processes in industrial markets. *European Journal of Marketing* 5:178–188.

Weick, K. E. (1976) Educational organizations as loosely coupled systems. *Administrative Science Quarterly* 21:1–19.

White, G. F. (1964) *Choice of Adjustments to Floods.* University of Chicago, Department of Geography Research Paper 93.

White, L. (1959) *The Evolution of Culture.* New York: McGraw–Hill.

Wicklund, R., and Brehm, J. (1976) *Perspectives on Cognitive Dissonance.* Hillsdale, N.J.: Lawrence Erlbaum Associates.

Wilbanks, T. J. (1979) Effective Social Science Research for Energy Policy. Paper presented to the American Association for the Advancement of Science, Houston, Texas.

Wilbanks, T. J. (1982) Decentralized energy planning and the potential of alternative energy systems. In J. Henderson, ed., *Environment, Energy, and Economics.* Proceedings of Centennial Conference, Tuskegee Institute, Tuskegee, Alabama.

Wilbanks, T. J. (1983) Energy self-sufficiency as an issue in regional and national development. In T. R. Lakshmanan, ed., *Energy and Regional Growth.* London: Gower.

Wildavsky, A., and Tenenbaum, E. (1981) *The Politics of Mistrust: Estimating American Oil and Gas Resources.* Beverly Hills: Sage.

Wilder, D. (1978) Effects of predictability on units of perception and attribution. *Personality and Social Psychology Bulletin* 4:604–607.

Williamson, O. E. (1964) *The Economics of Discretionary Behavior: Managerial Objectives in a Theory of the Firm.* Englewood Cliffs, N. J.: Prentice-Hall.

Williamson, O. E. (1975) *Markets and Hierarchies:* Analysis and Antitrust Implications. New York: Free Press.

Willrich, M. (1975) *Energy and World Politics.* New York: Free Press.

Wilson, J. (1966) Innovation in organizations: notes toward a theory. In Thompson, J. D., ed., *Approaches to Organizational Design.* Pittsburgh: University of Pittsburgh Press.

Wilson, J. P. (1976) Motivation, modeling, and altruism: a person x situation analysis. *Journal of Personality and Social Psychology* 34:1078–1086.

Winett, R. A., and Neale, M. S. (1979) Psychological framework for energy conservation in buildings: strategies, outcomes, directions. *Energy and Buildings* 2:101–116.

Winett, R. A., Love, S. Q., and Kidd, C. (1982) Effectiveness of an energy specialist and extension agents in promoting summer energy conservation by home visits. *Journal of Environmental Systems* 12:61–70.

Winett, R. A., Hatcher, J. W., Fort, T. R., Leckliter, J. N., Love, S. Q., Riley, A. W., and Fishback, J. F. (1982) The effects of videotape modeling and daily feedback on residential electricity conservation, home temperature and humidity, perceived comfort, and clothing worn: winter and summer. *Journal of Applied Behavior Analysis* 15:381–402.

Winkler, R. C., and Winett, R. A. (1982) Behavioral interventions in resource management: a systems approach based on behavioral economics. *American Psychologist* 37:421–435.

Wohlstetter, R. (1962) *Pearl Harbor.* Stanford, Calif.: Stanford University Press.

Wright, J. D., et al. (1979) *The Apathetic Politics of Natural Disasters.* Amherst, Mass.: Social and Demographic Research Institute, University of Massachusetts.

Yates, S. (1982) Using Prospect Theory to Create Persuasive Communications about Solar Water Heaters and Insulation. Unpublished doctoral dissertation. Santa Cruz: University of California.

Yates, S., and Aronson, E. (1983) A social-psychological perspective on energy conservation in residential buildings. *American Psychologist* 38:435–444.

Yergin, D. (1979) Conservation: the key energy source. In R. Stobaugh and D. Yergin, eds., *Energy Future.* New York: Random House.

Zaltman, G., Duncan, R., and Holbek, J. (1973) *Innovation and Organizations.* New York: Wiley Interscience.

Zimmerman, C. A. (1980) Energy for Travel: Household Travel Patterns by Life Cycle Stage. Paper presented at the International Conference on Consumer Behavior and Energy Policy, Banff, Alberta, Canada, September.

Appendix:
Biographical
Sketches of
Committee Members
and Staff

ELLIOT ARONSON is professor of psychology at the University of California, Santa Cruz. He previously taught at Harvard, the University of Minnesota, and the University of Texas. His recent work is in applying the techniques of experimental social psychology to specific social problems, such as ethnic prejudice and energy conservation. He has won distinguished research awards from the American Association for the Advancement of Science and the American Psychological Association, and his experiments in desegregated classrooms recently won the Gordon Allport prize for intergroup relations. Of the many books he has written or edited, *The Social Animal* received the National Media Award in 1972. He has a Ph.D. degree in social psychology from Stanford.

ROBERT AXELROD is professor in the Department of Political Science and the Institute of Public Policy Studies of the University of Michigan. His research interests include mathematical models of decision making and national security policy. He and biologist William Hamilton received the Newcomb Cleveland Prize of the American Association for the Advancement of Science for their article, "The Evolution of Cooperation." He received a B.A. degree from the University of Chicago and M.A. and Ph.D. degrees from Yale University.

ELLIS COSE was a National Research Council fellow who worked with the Committee on Behavioral and Social Aspects of Energy Consumption and Production. He is currently president of the Institute for Journalism Education at the University of California, Berkeley. Formerly, he was chief writer on the workplace and management for *USA Today,* an editorial columnist for the *Chicago Sun-Times,* a member of the editorial board of the *Detroit Free Press,* and a senior fellow and director of the Energy Policy Project at the Joint Center for Political Studies. He has been awarded several fellowships and has won several awards in journalism. He has an M.A. degree in science, technology, and public policy from George Washington University.

JOHN McCONNON DARLEY is professor and chairman of the Department of Psychology at Princeton University. He previously taught at New York University. His earlier work included research on people's reactions to emergencies, particularly reactions that determine whether people will give help to victims. Currently he studies perceptions of energy and energy problems and the ways in which information can be made available to people to facilitate their energy-conserving behaviors. He received a B.A. degree from Swarthmore College and a Ph.D. degree from Harvard University.

RICHARD HOFRICHTER worked as research associate for the Committee on Behavioral and Social Aspects of Energy Consumption and Production. A legal sociologist, he was previously research associate for the Criminal Justice and the Elderly Program of the National Council of Senior Citizens. He has written articles on neighborhood justice, legal services delivery, restitution, and victim compensation. His interests include democratic social planning, alternatives to courts, and class struggle in American cities. He received a Ph.D. degree in political science from the City University of New York.

SARA KIESLER is professor of social science and social psychology at Carnegie-Mellon University. She has been on the faculties of Yale University, Connecticut College, and the University of Kansas and served as a study director at the National Research Council from 1975 to 1979. Her current research interests include behavioral and social aspects of computing, telecommunications, technological change in organizations, group decision making, and public policy in the areas of technology, aging, and energy. She is an editor for *Social Issues* and a member of the Panel on Consensus Development of the National Institutes of Health, the executive council of the Society for the Psychological Study of Social Issues, and the council of the American Psychological Association. She received a B.S. degree from Simmons College, an M.A. degree from Stanford University, and a Ph.D. degree in psychology from Ohio State University.

DOROTHY LEONARD-BARTON is assistant professor of management at the Sloan School of Management of the Massachusetts Institute of

Technology. Her interest in the transfer of technologies began in Southeast Asia, where she spent eight years as a journalist. Subsequently she directed research at SRI International on the diffusion of energy-conserving practices and technologies. In her current research, she continues to study the psychological, sociological, and organizational factors influencing the acceptance or rejection of computer, medical, and energy-saving innovations. She has served on several advisory panels for the National Science Foundation and the Solar Energy Research Institute. She has a B.A. degree from Principia College and a Ph.D. degree in communication research from Stanford University.

JAMES G. MARCH is Fred H. Merrill Professor of Management and professor of political science and sociology at Stanford University and a senior fellow of the Hoover Institution. His research focuses on organizations, decision making, and leadership. He is a member of the National Academy of Sciences, the American Academy of Arts and Sciences, and the National Academy of Education. He received a B.A. degree from the University of Wisconsin and a Ph.D. in political science from Yale University.

JAMES N. MORGAN is a research scientist at the Institute for Social Research and a professor of economics at the University of Michigan. His major research, involving surveys on consumer behavior, has covered saving, income, and assets, the behavior of the affluent and of the poor, auto accident victims, and injured workers, and he has also conducted studies of philanthropy, retirement, and unpaid productive effort. He is the principal investigator of the Panel Study of Income Dynamics, which has been following a representative sample of individuals and their family situations for sixteen years. He is a member of the National Academy of Sciences and a fellow of the American Statistical Association and the Gerontological Society of America. He is also a longtime member of the board of directors of Consumers Union, a product-testing organization. He has a B.A. degree from Northwestern and a Ph.D. degree in economics from Harvard University.

PETER A. MORRISON is director of the Rand Corporation's Population Research Center in Santa Monica, California. Previously he was assistant professor of sociology at the University of Pennsylvania and a research associate at its Population Studies Center. His principal interests are applied demographic analysis and forecasting. His research has centered on a variety of policy-related topics, including urban and regional decline, the financing of Social Security, health care delivery, and personal energy use. He has served on the board of directors of the Population Association of America and on various committees of the National Science Foundation, the National Institute of Child Health and Human Development, and the Department of Agriculture. He received an A.B. degree from Dartmouth College and a Ph.D. degree in sociology from Brown University.

LINCOLN MOSES is a professor in the departments of statistics and preventive medicine at Stanford University. He has been Dean of Graduate Studies and Associate Dean of Humanities and Sciences at Stanford. He served as the first administrator of the Energy Information Administration, in the U.S. Department of Energy, from 1978 to 1980. His principal interests are in the applications of statistics to medical and behavioral research. He is a member of the Institute of Medicine and a fellow of the American Statistical Association and the Institute for Mathematical Statistics. He received A.B. and Ph.D. degrees from Stanford University.

LAURA NADER is professor of anthropology at the University of California, Berkeley. Her principal research interests have included law and social control, indirect controlling processes, the history and anthropology of science, and more recently, the behavioral components of energy research. She has worked in Mexico, the Middle East, and the United States and has produced two films on her research. She was recently a fellow at the Woodrow Wilson International Center for Scholars. She has a B.A. degree from Wells College and a Ph.D. degree from Radcliffe College (Harvard University).

STEVEN E. PERMUT is associate professor of organization and management and is affiliated with the Institution for Social and Policy Studies at Yale University. He is responsible for the graduate program in marketing management and consumer behavior and has particular interest in the applications of consumer behavior to problems of public policy. He is general editor of the Praeger Series in Public and Nonprofit Marketing and serves on the editorial boards of the *Journal of Marketing, Journal of Business Forecasting*, and *Journal of Consumer Marketing*, among others. He has an interdisciplinary Ph.D. degree in communications from the University of Illinois.

ALLAN SCHNAIBERG is professor of sociology and urban affairs at Northwestern University. He has done extensive research on sociopolitical conflicts over energy and environmental issues. He has served as a consultant to the Bureau of Land Management in its social effects study, as chairman of the environmental problems division of the Society for the Study of Social Problems, and is a member and frequent session organizer for the American Sociological Association. He has recently been engaged in international collaboration with the Institute for Environment and Society at Wissenschaftszentrum-Berlin. He is an associate editor for *Eval-*

uation Review. He has a B.S. degree in chemistry from McGill University and a Ph.D. degree in sociology from the University of Michigan.

ROBERT H. SOCOLOW is professor of mechanical and aerospace engineering and director of the Center for Energy and Environmental Studies at Princeton University. The goals of his current research are to clarify issues of science and values related to the global energy and environmental crisis. His specific interests include residential energy conservation strategies, technologies for economic development, and the role of analysis in decisions about natural resources. Since 1972, he has directed a team of scientists, engineers, architects, statisticians, and psychologists in a research program on energy conservation in the built environment. He received B.A., M.A., and Ph.D. degrees in physics from Harvard University.

PAUL C. STERN is study director for the Committee on Behavioral and Social Aspects of Energy Consumption and Production. He heads the energy committee of the Division of Population and Environmental Psychology of the American Psychological Association. Previously he was research associate in the Program on Energy and Behavior at Yale University's Institution for Social and Policy Studies. His research on energy and environmental policy issues has resulted in numerous articles and a book. He has a B.A. degree from Amherst College and a Ph.D. degree in psychology from Clark University.

THOMAS J. WILBANKS is associate director of the Energy Division at the Oak Ridge National Laboratory. He formerly served on the faculty at Syracuse University and as chair of the geography department at the University of Oklahoma, where he also participated in the Science and Public Policy Program. His principal interests are in energy policy, institutional roles and structures in science and technology, and relationships between society and technology. He has been a national officer of the Association of American Geographers and a member of a number of panels concerned with national energy research policy. He has a Ph.D. degree in geography from Syracuse University.

SIDNEY G. WINTER is professor of economics and of organization and management at Yale University. He was previously a professor at the University of Michigan, and has also taught at the University of California, Berkeley, and other universities. His research interests include behavior of the firm and industrial organization, with particular emphasis on technological change and the evolution of industrial structure. He received a B.A. degree from Swarthmore College and M.A. and Ph.D. degrees from Yale University, all in economics.

Index